Profits and Professions
Essays in Business and Professional Ethics

CONTEMPORARY ISSUES IN BIOMEDICINE, ETHICS, AND SOCIETY

Profits and Professions: *Essays in Business and Professional Ethics* edited by **Wade L. Robison, Michael S. Pritchard,** and **Joseph Ellin,** 1983

Visions of Women, edited by **Linda A. Bell,** *1983*

Ethics and Animals, edited by **Harlan B. Miller** and **William H. Williams,** *1983*

Who Decides?, edited by **Nora Bell,** *1982*

Medical Genetics Casebook, by **Colleen Clements,** *1982*

The Custom-Made Child?, edited by **Helen B. Holmes, Betty B. Hoskins, and Michael Gross,** *1981*

Birth Control and Controlling Birth, edited by **Helen B. Holmes, Betty B. Hoskins, and Michael Gross,** *1980*

Medical Responsibility, edited by **Wade L. Robison and Michael S. Pritchard,** *1979*

Contemporary Issues in Biomedical Ethics, edited by **John W. Davis, Barry Hoffmaster, and Sarah Shorten,** *1979*

Profits and Professions

Essays in Business and Professional Ethics

Edited By

Wade L. Robison
Kalamazoo College

Michael S. Pritchard

and

Joseph Ellin
Western Michigan University

Humana Press • Clifton, New Jersey

Library of Congress Cataloging in Publication Date

Main entry under title:

Profits and professions.

(Contemporary issues in biomedicine, ethics, and society)
Some of the papers presented at a conference and lecture series in professional and business ethics, jointly sponsored by the Departments of Philosophy of Kalamazoo College and Western Michigan University during the 1979–80 academic year.
Includes bibliographical references and index.
1. Business ethics—Congresses. 2. Professional ethics—Congresses. I. Robison, Wade L. II. Pritchard, Michael S. III. Ellin, Joseph. IV. Series.

HF5387.P76 1983 174 82-23399
ISBN 0-89603-039-3

Crescent Manor
PO Box 2148
Clifton, NJ 07015

Printed in the United States of America.

CONTENTS

Part I: Professional Ethics

Part II: Business Ethics

Part III: Professionals In A Corporate Setting

Preface

Suppose an accountant discovers evidence of shady practices while examining the books of a client. What should he or she do? Accountants have a professional obligation to respect the confidentiality of their clients' accounts. But, as an ordinary citizen, our accountant may feel that the authorities ought to be informed. Suppose a physician discovers that a patient, a bus driver, has a weak heart. If the patient continues bus driving even after being informed of the heart condition, should the physician inform the driver's company? Respect for patient confidentiality would say, no. But what if the driver should suffer a heart attack while on duty, causing an accident in which people are killed or seriously injured? Would the doctor bear some responsibility for these consequences?

Special obligations, such as those of confidentiality, apply to anyone in business or the professions. These obligations articulate, at least in part, what it is for someone to be, say, an accountant or a physician. Since these obligations are special, they raise a real possibility of conflict with the moral principles we usually accept outside of these special relationships in business and the professions.

These conflicts may become more accentuated for a professional who is also a corporate employee—a corporate attorney, an engineer working for a construction company, a nurse working as an employee of a hospital. For then, not only are there the conflicts that may occur between the requirements of ordinary morality and one's professional obligations and responsibilities, there is also the possibility of conflict between all these responsibilities and the legitimate expectations of employers.

The essays in this volume focus on issues involving professional ethics, business ethics, and the ethics of professionals in business. Parts I and II raise the question of whether business and the professions are—or ought to be—governed by principles and standards markedly different from those of ordinary morality. Do lawyers and business executives, for example, have a right, or even in certain circumstances a duty, to practice deception, ignore the interests of people adversely affected by their actions, or engage in other practices that ordinary morality condemns?

Part III brings together the issues raised in Parts I and II: The ethical problems of being a professional and those of being in business are compounded for the professional in business. Part III examines the special problems that professionals, following their professional ethics, might encounter as employees of a business organization. Consider engineers. To whom do they owe their main loyalty—the firm that employs them, the users of the products they design, or perhaps the taxpayer who often foots the bill? Since the vast majority of engineers work within a corporate setting, engineering is a profession that, perhaps more than any other, raises clearly the issue addressed in Part III: What special problems do professionals within the business world face? For this reason, we have included several essays that focus on ethical problems in the engineering profession.

Unlike other approaches, this book is not meant to be an attack on the current level of ethics in business and the professions. Taken as a whole, these essays constitute a constructive examination of the place of ethics in business and professional life. The essays struggle with the problems faced by those in business and the professions as they try to be responsible. The book is meant to appeal to people in business and the professions, as well as those in philosophy and academic ethics. But, we think the book should also be of interest to every thoughtful person who is concerned about the very important moral questions that the institutions of business and the professions inevitably raise.

Wade L. Robison
Michael S. Pritchard
Joseph Ellin

Acknowledgments

This volume had its inception at a Conference and Lecture Series in Professional and Business Ethics jointly sponsored by the Departments of Philosophy of Kalamazoo College and Western Michigan University during the 1979–80 academic year. We wish to thank the Michigan Council for the Humanities and the Franklin J. Machette Foundation for their financial support for our Conference.

Our volume is not a publication of the proceedings of the Conference. Only some of the papers presented at the Conference are included, and each of these has been substantially revised for this volume. Also, many of the papers included were not presented at the Conference. They were selected becaue of their special relevance to the themes of the volume.

Several of the essays in this volume have been published elsewhere. We wish to thank the publishers for permission to reprint these essays. Michael Bayles's "Professionals, Clients, and Others" is taken from his book, *Professional Ethics* (Belmont, California: Wadsworth Publishing Co.), 1981. A shorter version of Thomas L. Carson and Richard E. Wokutch's "The Moral Status of Bluffing and Deception in Business" appeared in *Westminster Institute Review,* Volume 1, April 1981. Norman Gillespie's "The Business of Ethics" appeared in *The University of Michigan Business Review,* November 1975. Gene James's "Whistle-Blowing: Its Nature and Justification" appeared in *Philosophy in Context,* Volume 10, 1980. A shorter version of Mike W. Martin's "Professional Autonomy and Employer's Authority" was published in *Ethical Problems in Engineering,* Al Flores, ed., 2nd edition, Volume 1 (Troy, NY: RPI Press), 1980. Gerald Postema's "Moral Responsibility in Professional Ethics" was published in *New York University Law Review,* Volume 55, April 1980.

Finally, we wish to give special thanks to Gwen West for her invaluable assistance in making arrangements for the Conference and in preparing the manuscript.

CONTRIBUTORS

MICHAEL D. BAYLES • *Westminster Institute for Ethics & Human Values, Westminster College, London, Canada*

THOMAS L. CARSON • *Department of Philosophy and Religion, Virginia Polytechnic Institute and State University, Blacksburg, Virginia*

RICHARD T. DE GEORGE • *Department of Philosophy, University of Kansas, Lawrence, Kansas.*

THOMAS DONALDSON • *Department of Philosophy, Loyola University of Chicago, Chicago, Illinois*

JOSEPH ELLIN • *Department of Philosophy, Western Michigan University, Kalamazoo, Michigan*

ALBERT FLORES • *Department of Philosophy, Rensselaer Polytechnic Institute, Troy, New York*

NORMAN CHASE GILLESPIE • *Department of Philosophy, Memphis State University, Memphis, Tennessee*

ROBERT V. HANNAFORD • *Department of Philosophy, Ripon College, Ripon, Wisconsin*

GENE G. JAMES • *Department of Philosophy, Memphis State University, Memphis, Tennessee*

KENNETH KIPNIS • *Department of Philosophy, University of Hawaii, Honolulu, Hawaii*

DONALD R. KOEHN • *Department of Philosophy, Illinois Wesleyan University, Bloomington, Illinois*

JOHN KULTGEN • *Department of Philosophy, University of Missouri, Columbia, Missouri*

ROBERT F. LADENSON • *Department of Philosophy, Illinois Institute of Technology, Chicago, Illinois*

MIKE W. MARTIN • *Department of Philosophy, Chapman College, Orange, California*

WILLIAM F. MAY • *The Kennedy Institute of Ethics, Georgetown University, Washington, DC*

LISA R. NEWTON • *Department of Philosophy, Fairfield University, Fairfield, Connecticut*

GERALD POSTEMA • *Department of Philosophy, University of North Carolina, Chapel Hill, North Carolina*

MICHAEL S. PRITCHARD • *Department of Philosophy, Western Michigan University, Kalamazoo, Michigan*

WADE L. ROBISON • *Department of Philosophy, Kalamazoo College, Kalamazoo, Michigan*

L. DUANE WILLARD • *Department of Philosophy, University of Nebraska, Omaha, Nebraska*

RICHARD E. WOKUTCH • *Department of Philosophy and Religion, Virginia Polytechnic Institute and State University, Blacksburg, Virginia*

Part I

Professional Ethics

Professional Responsibility

An Introduction

Wade L. Robison

In a recent case that acquired national prominence, a client told his lawyers that he had raped and killed two missing women. The lawyers visited the place where the bodies were hidden, but did not tell the authorities, or anyone else, for six months—despite being visited by the pleading parents of one of the young women. The lawyers held that the professional duty of confidentiality forbids disclosing the information until after the trial of their client. Clearly, many of us would find such a view shocking; as nonlawyers, we would, at the very least, feel obligated by our everyday morality to inform the inquiring parents that their daughter was dead.

An example such as this strikingly raises the issue of conflict between professional obligations and our ordinary moral ideas. This issue, with many of its important ramifications, is one of the two central problems discussed in the four essays that constitute Part I. If there are, as is often claimed, special standards governing professional conduct, why should there be? Professions sometimes claim to have high ideals, professionals having a "special calling." But our example suggests that professional morality is perhaps less admirable, less defensible, less based on ideals than ordinary morality. So we need to ask, "What are—and ought to be—the relationships between the standards and principles that govern the professions, and those that are endorsed by everyday morality?" When there is a conflict between professional and ordinary morality, does ordinary morality override? Or professional morality? Or is there a kind of irreducible conflict between the two?

However one settles that problem, there is another related problem of equal importance. Given the apparent conflict between professional and ordinary morality, a professional will often seem caught in the middle—forced to act as a professional in ways he or she would not act as an ordinary citizen. So one must be concerned with a second central problem, "How can professionals develop a coherent moral personality, with a coherent set of judgments, beliefs, and attitudes?" How can those lawyers not find their own behavior as shocking as we find it?

In the first selection, "Professional Responsibility and the Responsibility of Professions," Kenneth Kipnis explores concepts of responsibility that apply to the professions and the professionals within them. He develops a view of the nature of professions that implies that professions are obligated to serve the public good and that there are special, and in some sense higher, standards in the professions. Kipnis is concerned with the first of the two central problems we sketched, for he thinks that the fundamental philosophical question that needs to be asked about professions is: "What justification is there for having special standards?"

His answer is that since practitioners in a particular profession have control over who enters that profession and who does not, they have what amounts to a legal monopoly on the exercise of the skill for which they claim maximal competence. But, Kipnis argues, to have a legal monopoly is to have special privileges, backed by the state, and these privileges would lack point if there were no reasonable expectation that the members of the profession were committed to serving the public. Thus special standards are appropriate to assure that this reasonable expectation of public service is realized.

Kipnis notes that acknowledgement of the obligation of professions to serve the public is commonly made in the codes of ethics for the various professions. It is in these codes that one can expect to find articulated the special standards that apply to the professions and their practitioners. Since these codes attempt to delineate the public interest that is to be served by the various professions, Kipnis concludes that the codes are in need of rigorous philosophical scrutiny.

In "Professionalization: The Intractable Plurality of Values," Lisa Newton questions the extent to which the professions actually satisfy the view developed by Kipnis. According to Kipnis' view, a profession typically claims maximal competence in a given area of knowledge or skill and also claims a moral commitment to the public good. Newton adds that a moral commitment to the good of specific individuals served by professionals is a third feature of some professions and that a fourth feature, unmentioned by Kipnis, is that many professions are in a position to command large fees as a result of their legal monopoly over some service. She argues that these last two features of many professions may conflict with the ideal of serving the public good. In fact, she claims, a look at the historical development of the professions will show that it is not an essential part of the nature of a profession that it be morally committed to serving the public good as such.

Newton claims that professions actually develop in three different ways. A profession may come to exist, first, "for the sake of excellence in the practice" of some skill or some area of knowledge. Philosophers or cooks may become professionals to enhance their skills. Second, persons may become professionals by finding a way to market their skills or knowledge. A professional is someone who performs for a fee—a prostitute, or a professional musician. And, third, a person may become a professional out of a commitment to be of service to others. A person may become a nurse in order to help individuals.

These are three different routes to professional status, and none of them need involve any concern with the public good. A profession is marked, Newton claims, by there being an "identifiable skilled activity that constitutes professional practice" (with a means of transmitting the skill) and by the skilled activity being carried on for pay. If we examine actual professions like, e.g., being a philosopher, we discover these features, but no obvious appeal to the public good.

That appeal comes after the development of a profession, Newton claims. And it is best accounted for as a contract between the state and the members of a profession that the state will provide the members with a monopoly over a service if the members of the profession will act for the public good. "Public service," Newton concludes, "is the last professional commitment, not the first; it is the most doubtful of the profession's commitments, the one questioned most by its own members as well as by the society at large."

The essays by Kipnis and Newton raise fundamental questions about the relationship between the professions and the public good. Is it essential to the nature of a profession that it be morally committed to serve the public good? This is not an empirical question about what professions actually are like. It is, more fundamentally, a moral question about what professions ought to be like. And if professions are morally committed to serve the public good, is there a clear notion of the public good that could guide the professions? Even if such a notion is available, it can be asked how the professions might best serve the public good. As shall be seen in Part II, parallel questions can be raised about the responsibilities of those in business. For example, it has been argued that businesses have no direct social responsibilities. The public interest is best served, it may be claimed, by those in business seeking to maximize profits, subject only to the constraints of law. Should something similar be argued about the professions: The proper role of the professions is to serve their clients well—with the unintended consequence being that the public good is served? Or should the professions be morally committed to serving the public good as such?

However this issue is resolved, it is clear that, in actual practice, the lives of many professionals are beset with a tension between their personal moralities and what they take to be their professional responsibilities. In "Moral Responsibility in Professional Ethics," Gerald Postema takes up the issue of the relationship between personal and professional moralities and the nature of the moral personality that professionals take on when those moralities are at odds with each other.

There are, Postema claims, three different ways in which one might go about trying to solve the problem of conflicts between professional and personal moralities. One is to deny that there are special duties within any profession. But special duties are needed, he argues, if professions are to achieve the social goals we have designed them for, and there is nothing morally wrong with there being special duties, i.e., with our sometimes not acting on the balance of all reasons for doing something, but only on some, viz. those that are professionally relevant.

A second possibility would be to develop a "casuistry for a broad range of dilemmas." That is, one could develop specific rules for particular cases, perhaps generated, as in case law, by specific moral dilemmas. A third possibility would be to provide some general account of the relationship between the principles governing each morality so that one could solve particular problems as they arose. But these two possibilities would be of limited use in practice since they either consider isolated problems with no way of resolving new ones or require us to act from a single moral perspective when our difficulties stem from our having to assume several perspectives. In addition, and more important, they both rest on a mistaken view of moral judgment and moral experience. It is not enough for us simply to arrive at the morally correct solution. We must also have the appropriate attitudes and reactions since morality is "a matter of relating to people in a special and specifically human way." The focus of our concern ought therefore to be with responsibility as a virtue or trait of character.

The problem is that professionals may have difficulty maintaining integrated moral personalities: their professional roles require them to act in ways that are often directly counter to what they ought to do as private citizens. So their moral personalities may be either "fragmented or shrunk to fit the moral universe defined by the role."

A lawyer is a good example. The standard conception of a lawyer is that he or she is to be partisan for a client's interests and objectives and neutral with respect to the client's character or aims. But this conception requires a "sharp separation of private and professional morality." For then a lawyer must often be partisan for interests and objectives he or she would in normal circumstances object to strenuously and be neutral regarding character traits he or she would in normal circumstances deplore. The consequence is to adopt one or the other of what Postema calls "extreme strategies of identification." Either one takes on "an unquestioning acceptance of the duties and responsibilities of one's roles," or "one conforms only to avoid the external consequences of failing to do so, in no way internalizing the role or its basic principles." The former is a maximal identification with one's role, the latter a minimal identification. In either case, "practical deliberation, judgment, and action *within* the role are effectively cut off from ordinary moral beliefs, attitudes, feelings, and relationships—resources on which responsible judgment and action depend."

So a lawyer is encouraged by the standard conception of what a lawyer ought to do to cease to accept responsibility for his or her actions. One either identifies with one's role, and so shrinks one's moral personality, or separates oneself from one's role, and so fragments one's moral personality.

The proper solution, Postema claims, is to develop a unified conception of moral personality. He spends the remainder of the paper sketching a conception and answering objections. The main characteristics of that conception are that the professional must serve the functions of the role while "integrating his own sense of moral responsibility into the role itself."

This latter concern leads directly into the paper by Michael Bayles, "Professionals, Clients, and Others." For if a professional's own sense of moral responsibility is simply the fragmented or shrunken sense Postema discusses, it will do little good to integrate it into the role itself. The real concern is to find some sense of moral responsibility that is external to the ethics of one's own profession. Bayles raises this issue by asking, "Are professionals (e.g., lawyers) exempt from certain obligations?". That is, is a lawyer, for instance, exempt from the ordinary obligation to inform the authorities about a murder?

According to Bayles, Charles Fried is a good representative of the official position: "Acts performed in a representative capacity are subject only to legal constraints and not also moral ones." That is, one ought to fulfill one's obligations to clients over one's obligations to others as long as the conduct is legally permissible.

Bayles has a number of criticisms of this view. The most important is that Fried's position assumes that the legal system is just. For it assumes that there is no reason for assessing morally the acts one performs as a lawyer.

What Bayles suggests as an alternative to the official position is a general test: "adopt a consumer perspective and consider what balance would most promote the values of a democratic society." The test is not a simple one, and Bayles considers a number of steps necessary for the procedure and several examples. One will give the flavor of the test: if a lawyer discovers that a client is going to commit perjury, ought he or she remain silent or inform the judge? It has been argued that the lawyer's responsibility of confidentiality is more important than the public's interest in convicting criminals and avoiding perjury. Bayles argues that by his proposed test a lawyer should inform. He claims that no clear reason exists for a reasonable person to prefer that a lawyer not inform, and so knowingly help a client lie to the court, rather than that a lawyer inform.

We thus end this section of the volume with a possible answer to one of the questions with which we began: What relationships are there between standards and principles of the professions and ordinary moral standards and principles? Bayles is arguing that any conflict between the two sorts of standards and principles ought to be settled by appealing to what ordinary reasonable persons would say.

It should be emphasized that, however plausible, this is but one possible answer to the question. We ought to tread carefully in accepting suggested solutions. For one thing, in pursuing either of the two central problems with which we began, one quickly finds oneself involved in difficult and complex questions, all of which need to be sorted out and clarified before one can even begin to assess how a suggested solution affects them. The disputes between Newton and Kipnis are illustrative of that. For another, there is an interdependence that cannot be ignored. A suggestion regarding the nature of a profession, for instance, has ramifications regarding how one ought to act as a professional and a nonprofessional, while a suggestion regarding the nature of an independent standard may have ramifications regarding the nature of a profession. One must assess any proposed suggestions in light of their possible implications for all the issues raised.

Professional Responsibility and the Responsibility Of Professions

Kenneth Kipnis

If one studies the statements emerging from those organizations that undertake to speak for professions, one is struck by the codes of ethics and canons of professional responsibility that appear so frequently as to make them seem almost the hallmark of professionalism itself.[1] These codes appear to be based on the assumption that some actions can merit one assessment if undertaken by a certain professional, but another assessment if undertaken by some other person. For a philosopher, perhaps the most interesting thing about professions is their suggestion that there somehow exist certain special justifiable standards for the conduct of a certain class of persons.[2] At least two questions are raised. First, one wonders what might be the justification for these special standards?[3] And second, one wants to know how it is possible to delineate the proper dimensions of these special standards and to assess the magnitude of the claim that they have upon professionals. These are, I submit, the two most important philosophical issues that arise in the field of professional ethics. It is the aim of the present essay to map an approach to them.

The argument here presented attempts to show how distinctive responsibilities of the individual professional can be understood in terms of the responsibility of his or her profession, taken as a whole. In the first section, I will be developing a concept of responsibility that is fundamental in this context. In the second, I will show how professions can come to have such responsibilities. Finally, I will sketch the relationship between the responsibilities of professions and the responsibilities of professionals.

I

A place to begin is with the idea of having a responsibility.[4] A married couple can have the responsibility for the welfare of their child. A philosophy instructor can have the responsibility for a particular course. A member of a mountaineering expedition can have the responsibility for the maps. The sanitation company can have the responsibility for the proper collection and disposal of a community's garbage. In these cases and others, what one has responsibility *for* is always a matter of some concern. One would have to tell a story to explain how it is that someone has responsibility, say, for a pebble; why, for example, it is a pebble of considerable importance; that it is, in fact, a crucial piece of evidence in a criminal case. In general, the more important the matter of concern, the greater the responsibility.

Those who have responsibility for matters of concern are expected (in the evaluative—not the predictive—sense of "expected") to give these matters all the attention they are due. They are expected to be able to provide—and can generally be called upon to provide—an account of how they are insuring that due care and attention are being given to the matters of concern for which they have responsibility. When we talk abut a responsibility that an agent has, we are referring to a sphere of concern for which the agent is accountable. If a preventable mishap occurs within the sphere, there is a strong presumption that the agent is blameworthy. This concept of a responsibility—a sphere of concern for which an agent is accountable—is the concept I am taking to be fundamental here. To distinguish it from two derivative concepts of responsibility to be sketched shortly I will refer to it as "substantive responsibility."

Not every agent is well-suited to assume substantive responsibilites. A derivative concept of responsibility—what I will call "virtue responsibility"—is applied to persons who are, in a particular way, well-suited to assume important substantive responsibilities. Of the responsible person Graham Haydon has written:

> He will . . . be aware that he stands in, or has the opportunity of entering into, certain relationships with certain others in such a way that he in particular can be called to give an account of what happens in certain specified spheres. But in recognizing that he can be called to account for the nature and outcome of his actions, he is recognizing . . . the importance of so acting as to be able to give a good account.[5]

Thus the responsible person will be one who takes seriously the act of assuming a substantive responsibility and, having assumed it, is predisposed to give that responsibility the care and attention it is due.

Conversely, the person who is irresponsible is one who is predisposed to carelessness and/or inattention to his or her responsibilities. It is not difficult to understand why irresponsibility should be regarded as a vice. Many of our activities are undertaken within the contexts of relationships and organizations requiring a high degree of mutual reliance. The settling of responsibilities within the relationship or organization is often necessary if the actions of the participants are to make sense. Consider two pilots flying a large aircraft over the sea. If each does her or his part, attending to the need for coordination, progress can be made in safety. But either can, through inattention, frustrate the good-faith efforts of the other, endanger passengers, and, in general, erode the trust that makes cooperation possible in the first place. If we have determined that someone is irresponsible, we typically relieve that person of at least some important responsibilities and refrain from delegating new ones.

A second derivative sense of responsibility should be noted. If the child suffers from the neglect of its parents, if the philosophy course serves only to convince students of the pointlessness of the discipline, if the maps for the mountaineering expedition are damaged or misplaced, we may hold certain persons to have responsibility for these mishaps. What I would like to call "mishap responsibility" can occur in two ways: when someone acts so as to bring the mishap about and when someone with substantive responsibility for a matter of concern neglects to give due care or attention and, as a result, the mishap occurs. When mishaps occur, those with substantive responsibility for relevant matters of concern are typically called upon to provide accounts of themselves.

It appears to be a fact of contemporary life that concerns about irresponsibility will generally be related to whether or not a mishap has occurred and, in addition, to the scale of the mishap. The first major nuclear accidents have provoked a most penetrating interest. However, where the irresponsibility is clear and the risk of serious loss high, then, even if no mishap results, concern may well be great. If a flight crew throws a drunken party while aloft in a loaded aircraft, interested persons may well take a hard look at the personnel and the procedures of the airline in question.

A word needs to be said about the way in which one comes to have responsibilities. In the clearest cases, responsibilities are more or less explicitly settled by the parties themselves or by authorized representatives.[6] There are matters of concern that must receive attention and there is a collection of persons who create or maintain a system of relationships—an organization or organized activity—such that the responsibilities and the means of meeting them are allocated in a reasonable manner. The process of settling responsibilities is an ongoing political process. There are often questions about the privileges, rights, resources, immunities, and status accorded to those with responsibilities. These concomitant benefits can be a significant incentive to assume important responsibilities and can also be forfeited upon being relieved of responsibilities. Accordingly, disagreement about the best way to attend to matters of common concern may mask conflict about whose special interests will receive the most abundant satisfaction. Beyond the issue of benefits, there are questions about the standards to be met in attending to responsibilities and, moreover, questions about the quality of past performance. Finally, there are questions about how responsibilities ought to be delegated; whether, for example, those who have them ought to be relieved of them.

Consider, for example, the responsibility that parents have for their children. It is profoundly important that children be cared for and the institution of the family is the prevailing means by which this care is expressed. The legal and social context of parenthood accords privileges, rights, resources, immunities, and status to persons who are parents in the expectation that in that manner the welfare of children can be insured.[7] As things now stand it is up to parents, first of all, to insure that their children can mature into adults in a wholesome way. A failure here may well be a moral failure, a failure to meet one's responsibilities in a context where significant social values are at stake. In cases of such failure, where parents are shown to be unfit for the task of childrearing, they are relieved of their responsibilities by the courts.

Consider the member of the mountaineering expedition with responsibility for the maps. Just as the others rely upon that individual to attend to this important matter, the map carrier can rely upon the others to attend to a broad range of other concerns. If, however, it turns out that the map carrier cannot be relied upon, prudence requires that he or she be relieved of responsibilities in this context. Here, as with parenthood, responsibilities are more or less explicitly delegated. There is a reliance upon the person with responsibility, a granting of benefits in recognition of that responsibility, and the option of relieving the agent of that responsibility in the event that the arrangement proves unsatisfactory.

If one thinks of substantive responsibility in this general way, organizations as well as persons may have responsibilities. The sanitation company can, through authorized representatives, enter into an agreement with an authorized representative of a second organization, an incorporated township. If the agreement provides that the sanitation company shall have the *exclusive* right to pick up and dispose of the garbage, and that the township shall provide consideration in return for these services, then it seems unproblematic that the company has the responsibility for picking up all of the garbage in the township.

But now suppose that the garbage does not get picked up—let us say the company discovers that it does not have the resources to meet its responsibilities under the terms of the agreement. Can we make sense out of the suggestion that the company itself—as opposed to particular persons working within it—might be "irresponsible"?[8] I believe we can. To be sure, there clearly are cases where the failure of an organization to meet its responsibilities is a consequence of the failure some person within the organization to meet his or her responsibilities. But the matter cannot rest here. For an organization is more than a collection of persons: it is a system of positions and associated procedures. Thus, if an irresponsible person occupies a position and, as a consequence, a mishap occurs, it may be that the *procedures* used to select persons to occupy positions of responsibility *are themselves defective* in some way. Moreover, it may be that in some cases each person within the organization can meet his or her responsibilities and yet, *because the structure of the organization itself is faulty, its responsibility is not met.*[9] Perhaps the structure impedes the flow of certain types of information. Perhaps there are important matters that are not assigned as anyone's responsibilities. Perhaps the organization trains its personnel to attend only to some aspects of a problem, neglecting others; or only takes on persons who do; or singles them out for promotion, eventually discarding the others. One can reply that in these cases the charge of irresponsibility should lie with those persons who have designed the structure of the organization. But organizations rarely have such architects. Rather, they are generally aggregations of positions and procedures that have accumulated gradually over the years. (Such structures can lack coherent justifications. We can only explain them as the products of historical processes.) And even where a defect-ridden organization can point to its architect, one should still ask how the responsibility for design happened to settle on a person so deficient in organizational competence.

Although we may not be comfortable talking about organizations as "irresponsible," there can be no doubt that some of them are predisposed to carelessness and inattention toward the responsibilities they have assumed. Social institutions are artifacts, instruments that people have fashioned to realize human values. As with any artifact, they can be fashioned with the care and skill they require or they can be object lessons in negligence, irresponsibility, or even viciousness. If one is concerned with the propensity of a particular organization to fail to meet its responsibilities one cannot be blind to the structural aspects of the problem. In a society that routinely delegates to many organizations responsibility for a broad range of matters of public concern, close attention should be given to the manner in which persons within those organizations are selected, trained, promoted to positions of authority, and constrained to organize themselves so as to give due attention and care to our concerns. In the following section I will show how it is the delegation of such responsibility that creates a profession.

II

The development of an occupation into a profession is a lengthy process.[10] Occupations may be located at virtually any point along a continuum with day laborers at one end and doctors and lawyers at the other. Nevertheless, three distinct but related features characteristically emerge as professionalism progresses.

1. In the first place, organized practitioners within an occupation begin to make a *claim to maximal competence*. Through representatives, one begins to hear that a certain discrete class of persons, through intense and thorough intellectual training and practical experience, exceeds in skill all other persons in some area of endeavor. Two conditions must be satisfied before such a claim can be made. First, there must be some organization of practitioners within the favored class. The organization need not include all of them, but must include enough to warrant a claim to speak on behalf of the whole class. Second, there must be some criterion for deciding who belongs to the class of favored practitioners and who does not. (In the end, this evolves into an elaborate gatekeeping procedure involving education, accreditation of schools, and certification of new members.) The esoteric nature of the knowledge and skill possessed by this select class of practitioners begins to make it look as if persons outside the favored class simply lack the standing to make sound judgments about the performance of those specialists. As the claim to maximal competence comes to be accepted, it begins to seem more and more reasonable to let the select class of practitioners certify and evaluate itself. The organization of favored practitioners stands ready to assume this responsibility.

2. The special competence possessed by members of a profession is of a sort that can be applied to some matter of public concern. For reasons that will become clear, the process of professionalization requires that the profession make a public *commitment to devote itself to the realization of some significant social value,*[11] to give due attention to the distinctive matter of social concern to which its skills and knowledge can be directed. Typically one finds explicit reference to these values in the preambles to the codes of ethics that emerge from professional organizations. Thus, the American Bar Association's *Code of Professional Responsibility* begins with the following:

> The continued existence of a free and democratic society depends upon recognition of the concept that justice is based upon the rule of law grounded in respect for the dignity of the individual and his capacity through reason for enlightened self-government. Law so grounded makes justice possible, for only through such law does the dignity of the individual attain respect and protection. Without it, individual rights become subject to unrestrained power, respect for law is destroyed, and rational self-government is impossible.
>
> Lawyers, as guardians of the law, play a vital role in the preservation of society. The fulfillment of this role requires an understanding by lawyers of their relationship with and functions in our legal system. A consequent obligation of lawyers is to maintain the highest standards of ethical conduct.[12]

The same type of self-identification can be seen in official statements from professions of less lofty pretension. Consider these words from the preamble to the *Code of Ethics* of the National Association of Real Estate Boards:

> Under all is the land. Upon its wise utilization and widely allocated ownership depend the survival and growth of free institutions and of our civilization. The Realtor is the instrumentality through which the land resource of the nation reaches its highest use and through which land ownership attains its widest distribution. He is the creator of homes, a builder of cities, a developer of industries and productive farms.
>
> Such functions impose obligations beyond those of ordinary commerce. They impose grave social responsibilities to which the Realtor should dedicate himself, and for which he should be diligent in preparing himself. The Realtor, therefore, is zealous to maintain and improve the standards of his calling and shares with his fellow-Realtors a common responsibility for its integrity and honor.[13]

These examples can be multiplied. It is only rarely, however, that professional associations precisely delimit the social values to which their skills are characteristically to be directed and to which they pledge to commit their energies. Nonetheless, what is clear is that professions typically endeavor to represent themselves as dedicated to an ideal of social service.[14]

3. Neither the competence nor the commitment are sufficient to change a discrete organized occupational activity into a profession. What is needed is social recognition of and reliance upon the organized profession as the means by which certain skills are to be applied and certain matters of public concern given the attention they are due. This last feature is built upon the preceding two. For if there is confidence that the members of a profession possess maximal competence in a particular area and if there is trust that these practitioners are deeply committed to the responsible application of their distinctive skills, then there will seem to be neither the ability nor the need to designate nonprofessionals as overseers of professional practice. No one is competent to do the job and we do not need to have it done in the first place.

As this view becomes more plausible, as the profession earns or otherwise secures its trust and confidence, it gradually obtains autonomy at both the institutional and individual levels. Institutional autonomy is secured when the organization obtains recognition as representing the profession. It is invited to speak on behalf of the profession. It takes control over membership: the selection and training of candidates, the accreditation of professional schools and programs, the certification of new members, the promulgation and enforcement of standards of professional practice. In the clearest cases of professionalization, the profession becomes, in essence, an unregulated legal monopoly with respect to a certain service, unauthorized practice being a criminal offense. Individual autonomy is secured when the profession obtains a substantial measure of control over the conditions and content of its work. It lays claim to, and is granted, the latitude it says it needs in order to do its best job. Autonomy for the individuals in a profession can take several forms. The tradition of academic freedom is probably the best known mechanism, but others can be seen in the governance structures of hospitals[15] and in the operation of Disciplinary Rule 3 of the ABA code, which effectively prevents lawyers from working for nonlawyers in corporations that sell legal services. Substantial control over the conditions and content of work is, in each context, secured by members of the affected profession.

In the end, of course, it is we who, through our representatives, delegate responsibility to professions or relieve them of it. With respect to professions as a whole and the communities they serve, responsibilities are settled, in the clearest cases, in committees of state legislatures. These committees typically meet with the representatives of professional organizations in efforts to hammer out satisfactory arrangements. I have not been convinced that the delegation of such responsibility is always unwise.[16] Much, it seems, depends upon the political dimensions of the institutional and individual autonomy that is granted, the soundness of the profession's claim to competence (the scope of that competence relative to the epistemic legitimacy of the knowledge-claims upon which it is based), and the degree to which the profession is genuinely committed to an ideal of public service. A complete justification of a profession would have to take account of all of these matters. Here, however, what is of most interest is the commitment. For it assures us that the privileges that have been granted will not be abused or otherwise exercised irresponsibly. If the profession plainly has organized itself in such a way as to insure that its distinctive matter of social concern receives from the profession the attention it is due, then it may well be reasonable for us to choose to maintain a system of professional practice. But how might one assess the manner in which a profession meets or fails to meet this condition?

III

A place to begin is with the code of ethics of the profession. These codes embody a theory about the relationship between the profession and the community it serves: lawyers, it will be remembered, are "guardians of the law." The codes indicate in their preambles the significant social values that the professions take to be their primary concerns. Although there is often a pressing need for a precision that philosophical skills could provide, the central intent is clear: to register, on the part of the profession, an official recognition of the sphere of responsibility to which the profession owes due attention.

But how, we would want to know, is this recognition of responsibility—the reciprocal of social privilege—to be acted upon by the organized profession? One way is through the institutionalization of a code of ethics, a set of principles defining the conditions under which professional skills are to be applied. Naturally, before we were through, we would want to consider the mechanisms by which a profession could insure that its members were adhering to its principles. These might or might not be enforcement procedures of some kind but, in any case, we would want to know that they were adequate to provide substantial adherence to the code of ethics under the existing conditions of professional practice. For a justifiable professional code consists of principles that would, if substantially adhered to, insure that the substantive responsibility of the profession receives the attention it is due. The rationality of the delegation of responsibility to professions can thus turn on the soundness of those principles.

Consider, for example, the duties that arise when someone makes it clear that he or she is committing a serious wrongdoing that is punishable by law. Assume that this takes place in a situation in which one is expected not to divulge the information. Should one let the proper authorities know? In nonprofessional contexts one can envision the arguments going either way. But if it had been a doctor who received the information while acting in a professional capacity, then the following argument would have application. The principle of doctor–patient confidentiality is one of the best-known and most firmly established principles in medical ethics. It is a part of the Hippocratic oath and its special claims upon the medical profession are recognized in legal privilege.[17] The justification for this principle is reasonably clear. As things stand, the medical profession has a special responsibility for the health of the community it serves. Persons who are not medical professionals usually lack the skill and confidence that doctors have and, even if they were competent, they are barred by law from putting their knowledge and skill into practice. The granting of special privileges to the medical profession would lack point in the absence of a reasonable expectation that the members of the profession were undertaking to use those privileges to serve (rather than to exploit) the public. Doctors have indeed been quite successful in their efforts to encourage the public to rely exclusively upon their profession in attending to certain concerns. For the medical profession to meet its responsibility, the public must be willing to see doctors when the need arises and to supply information that doctors need if they are to do their best work. It is reasonable to suppose that if doctors—even a very small number of doctors—are willing to use that information to the detriment of the patients who supply it, the public will be less willing to cooperate with doctors. A betrayal of confidence by a medical professional can erode public trust and poison the environment in which medical practice flourishes. On the other hand, adherence to the principle of doctor–patient confidentiality can help to create an environment in which the medical profession can do its best job.

Construed in this general way, the special responsibilities of individual professionals are based upon the responsibilities that their professions have assumed in the political process of professionalization. And, if this account is apposite, the principles of professional ethics—some of them at least—can be justified as sets of rules, substantial adherence to which will conduce to the creation of an environment in which due care and attention are best given to the distinctive matters of concern for which professions can be maintained.

It is important to note here that small changes in the articulated commitment of the profession can produce great changes in the code of ethics. Thus some disputes about what appear to be ethical issues may mirror deep differences concerning what the commitment of the profession ought to be. Consider, for example, the obligations that arise when a doctor believes that the small child being treated has been battered by the parent who brought it in. Assume that, although it is possible for the doctor to save this child from further harm by notifying the proper authorities, if the doctor does this, other child abusers will hesitate before taking their injured children to doctors, with consequent fatal results for these other children. Still other parents will hesitate out of a fear that they will be wrongly identified as child abusers upon bringing their injured children to the doctor. Now on one account, the doctor ought to be the champion of the abused patient. The medical profession should assert its overriding obligation to the patients that are delivered into its hands. Nothing can be more important than the suffering patient that is before the doctor and the interest that that patient has in securing what it can of physical well-being. Yet on another account, the medical profession is the means by which we express our concern for our physical well-being. The medical profession that does the best job is the one that best secures the health of the community it serves. Thus, although at the outset, some children may be better helped by notifying the authorities, in the long run a medical profession that routinely betrays these parents will lose the trust it needs if it is to do its best work. Nothing can be more important to a doctor than being worthy of trust. On the one account the medical profession is primarily concerned with the health of the patients that are brought to it. On the other account the concern is with the general physical well-being of the community. A resolution of the ambiguities in the present codes may press hard choices upon the professions and their membership and upon those who delegate responsibilities to them.

Although there are other matters to be considered, it is the codes that merit the closest attention in the philosophical study of professional ethics. They delineate the public interest that the profession exists to serve. They can set out principles that, if substantially adhered to, will ensure that the public interest will receive from the profession the care and attention it is due. The claim that sound principles have upon the practitioner has its basis in the claim that the profession has to the trust of the public. A responsible profession, one might say, is one that does not merely profess its dedication to the public interest in its code but, rather, expresses that dedication in its very constitution.

From the point of view of the individual professional, the theory here presented calls attention to the political role that the practitioner can properly play within the profession itself. As a member of a profession, as one who can participate in its governance, he or she shares responsibility for the quality of its structure. If that structure merely serves to create the illusion of concerned attention, if it serves values that are quite different from—or even at odds with—the values the profession claims to serve, then the need is great for effective political action within the profession.[18]

From the point of view of the citizen, the theory here presented calls attention to the power to delegate responsibility to professional associations and the power to relieve them of responsibility. Deprofessionalization, the process of relieving professions of their responsibilities (and concomitant privileges and benefits), has been urged by some critics.[19] At a minimum, prudence requires that one refrain from leaving important matters to organizations that are predisposed to inattention as regards the responsibilities they have assumed. The code of ethics of a profession is a guarantee that the practice of individual professionals will conform to principles that display due regard for the profession's sphere of responsibility. Given the extent of our reliance upon the professions, it would not be unreasonable to subject these guarantees to the most rigorous philosophical scrutiny.

Acknowledgments

This paper was written in part during the NEH-sponsored National Project on Philosophy and Engineering Ethics at Rensselaer Polytechnic Institute. I am indebted to Sara Lyn Smith, Andrew Jameton, Haskell Fain, Michael Martin, Albert Flores, Lisa Newton, and Robert E. Ladenson for helpful comments and suggestions.

NOTES AND REFERENCES

[1]Most of these are to be found in Jane Clapp's *Professional Ethics and Insignia* (Metuchen: Scarecrow Press, 1974).

[2]Allison Jagger has considered philosophy itself as a profession in "Philosophy as a Profession" in *Metaphilosophy* 6 (1975): 100.

[3]At least one commentator has denied that there are such special standards. See Robert M. Veatch, "Medical Ethics: Professional or Universal," *Harvard Theological Review* 65 (1972): 558. Veatch's main argument has been criticized in Kenneth Kipnis's "Professional Ethics" in *Business and Professional Ethics* 2 (1978): 2. Recently, Veatch has modified his views in "Professional Medical Ethics: The Grounding of Its Principles," *Journal of Medicine and Philosophy* 1 (1979): 1.

[4]The discussion here builds upon H. L. A. Hart's analysis of "role responsibility" developed in Chapter IX of his *Punishment and Responsibility* (New York: Oxford University Press, 1968).

[5]Graham Haydon, "On Being Responsible," *Philosophical Quarterly* 28 (1978), 56.

[6]There is another view in which responsibility is generated where the serious need for attention encounters the capacity to meet that need. Thus, on this account, if someone should discover that a serious mishap can be averted through attention that only he or she can provide, then, even in the absence of any explicit delegation or assumption of responsibility, that person can have a substantive responsibility for the matter of concern involving the imminent mishap. Some of the dimensions of this theory are brought out in Kenneth Kipnis, *Philosophical Issues in Law: Cases and Materials* (Englewood Cliffs: Prentice-Hall, 1977), 99–158. See also the bibliographical references on pages 328–330. Though I am in some sympathy with those who advance this view, I have not been convinced by the arguments that have been brought forward in defense of it.

[7]Note that even if this expectation were unfounded, even if it were conclusively shown that other means were better suited to the task of rearing children, even if it were demonstrated that parents should not have the special rights and responsibilities that they presently have, parents would still have those rights and responsibilities, both legal and moral. They are, I would say, built into the present social structure and cannot be altered until that structure is changed.

[8]The issues surrounding organizational irresponsibility are helpfully represented in Christopher D. Stone's *Where the Law Ends* (New York: Harper and Row, 1975).

[9]The general issue of collective, nondistributive responsibility is treated in Peter A. French, ed., *Individual and Collective Responsibility* (Cambridge: Schenkman, 1972). See especially the articles by R. S. Downie and David Cooper.

[10]See Harold Wilensky, "The Professionalization of Everyone," *The American Journal of Sociology* 70 (1964), 137.

[11]See Talcott Parson, "Professions," *International Encyclopedia of the Social Sciences* 12 (1968), 536.

[12]Quoted in Maynard E. Pirsig, ed., *Cases and Materials on Professional Responsibility* (St. Paul: West Publishing Co., 1970), 337.

[13]Quoted in Stephen J. Martin, ed., *Commentaries on the Code of Ethics* (Bloomington: Indiana University Press, 1974), v. I am grateful to Mary Carson Smith for calling this material to my attention.

[14]In the end this produces a distinctive type of transaction in the payment of fees to professionals. They are paid, not for their work, but, rather, in order that they may do their work. Lawrence Haworth develops this point in his *Decadence and Objectivity* (Toronto: University of Toronto Press, 1978), 112.

[15]I am grateful to Andrew Jameton for calling the matter of hospital governance to my attention.

[16]Milton Friedman presents a strong case in "Occupational Licensure" in *Capitalism and Freedom* (Chicago: University of Chicago Press, 1963).

[17]See, for example, Rule 27 of the Uniform Rules of Evidence.

[18]There may be substantial obstacles in the way of effective political action within the profession itself. The structure of many professions precludes significant participation by large and distinct segments of its membership: the AAUP and nontenured professors, the American Bar Association and sole practitioners, the various engineering professional associations and working (nonmanagerial) engineers.

[19]See, for example, the writings of Ivan Illich: *Deschooling Society* (New York: Harper and Row, 1972); *Tools for Conviviality* (New York: Harper and Row, 1973); and *Medical Nemesis* (New York: Pantheon Books, 1976).

Professionalization

The Intractable Plurality of Values[1]

Lisa H. Newton

Any account of the professions, and of the role played by professions in the larger society, must meet two criteria: it must include the salient characteristics of professions, even the characteristics apparently in conflict; and it must make logical and moral sense of the substantial rights and responsibilities assigned to them. The two features most publicly associated with professionalism, by any profession's own declaration at least, are maximal competence in a certain area of knowledge or skill, and a moral commitment to the public good in that area; these features may be treated as the individually necessary and jointly sufficient conditions to justify "professionalization," i.e., *the social award of a legitimate monopoly of practice in that area to the organized profession*. A third prominent feature of some professions is the commitment on the part of individual professionals to the welfare and interests of individuals in their charge—clients, patients, students, or parishioners—even when (as in certain cases of maintenance of "confidentiality") protection of such interests is contrary to the public good and thus apparently derogates from the "public service" commitment above. This feature too can be integrated into the larger picture of the common good as productive of social welfare in the long run. By reassuring the public that confidences will be respected, such a provision increases public trust in the profession and encourages the public to resort to professionals in time of need. There is also a fourth characteristic of most professions: the professionals seem to be able to command very large fees. This feature can be justified as a benefit of legal monopoly. For example, remuneration is one more inducement to the profession to adhere to the set of rules identified as its "professional ethic," general adherence to which will benefit the society as a whole: if the ethic is substantially violated the monopoly will be withdrawn and the benefit will be lost.[2] The first criterion is well met by this account; the second is satisfied in the assumption that the specific moral commitment of any profession fits an area of social concern or need: there must be a match between the service that the profession is qualified to provide and an area of public need that would be very difficult to provide for through ordinary public facilities. Such a close fit between professionally offered service and demand—conjoined with the requirements that provision of the service entail lengthy training in esoteric skill and knowledge, that the providers offering the service can validate claims to competence in it, and that they have publicly committed themselves to use their skills only for the public good—justifies entrusting that area of social need to the providers of that service.

Given the characteristics of a profession, and the account that rationalizes them, it would seem to follow immediately that a profession could be held responsible for professional practice only as a whole, i.e., collectively. For professionalization, social legitimation of monopoly of practice in the area of maximal competence, is certainly an assignment of responsibility in any of Kurt Baier's[3] senses of "responsibility": it is an assigned *task* in an area of concern, for performance of which the profession is henceforth publicly *accountable*, mishaps in which will indicate professional *culpability*, and for which mishaps the professionals may very well be held *liable*. But to whom, or what, is the professional *answerable*, to complete Baier's list of senses of "responsible"? Not to the individual client or patient, by definition: the designation of a profession implies the incompetence of the layman to judge professional performance. The only superior to which a professional could be answerable is the state itself that granted professional status, and he or she is answerable there only through the profession as a whole, to which that status was actually granted. (The state, of course, has no more professional competence than any layman; it can call the profession to account not by virtue of superior expertise, but by virtue of its inalienable sovereignty over all areas of social concern. The state simply cannot assign any area of public action, welfare, or responsibility to any private group irrevocably; the consent of the people to the government cannot be construed to permit permanent loss of popular sovereignty in any category. Ultimately, the state *must* say whether the public is being properly served, so all who claim to serve the public must report to it.) It would follow that the professional must consider himself primarily responsible, not to the desires of his clients nor to the immediate demands of the larger society, but to his own professional ethic. It is the function of that ethic to spell out the obligation of the profession as a whole to the society as a whole—and it is the philosophers' job, perhaps, as members of that society particularly well-trained for such things, to keep these "professional ethics" under careful surveillance, to ensure that they do in fact serve the public interest.

In some respects, however, this account fails to explain certain attitudes and current practices of professionals, also certain of our attitudes and current practices with regard to them. Paying attention to these areas of discrepancy may serve to illuminate other possibilities for coming to grips with a surprisingly complex social phenomenon. To begin with the last point, it should be noted that we do not in fact hold professionals *collectively* accountable for professional performance, nor is it clear that we could if we wanted to. In fact we permit laymen to bring lawsuits against professionals alleging malpractice—alleging, that is, professional incompetence, as opposed to simple negligence. To be sure "expert testimony," support from other professionals of the same sort, is solicited by the plaintiff's lawyers in such cases; still, the suit originates with a layman's flat-out judgment that the professional did not know what he or she was doing, and terminates with a lay jury's verdict on that judgment. Both would be inappropriate in a context where the delegation of social responsibility, including the award of the monopoly of practice, had been to a profession collectively, and where that delegation has been justified by acknowledgment of a monopoly of competence in the field. The completion of "professionalization" would entail legal restraints on such private actions, permitting only professional review of complaints of inadequacy of professional performance. What would be the effect of such unprecedented legal restraints on the law, specifically upon the legal presumption of individual accountability for individual action? What would be the effect of establishing such unprecedented legal immunities, especially for groups already under suspicion of overprivilege, upon the already fragile morale of an envious and conflict-ridden populace? Would any attempt to secure sole jurisdiction of professional review boards be well-advised in this era of consumerism?

Second, there is the difficulty that professional organizations have at present only very limited authority in their own fields of practice. I do not know how many practicing philosophy professors belong to the American Philosophy Association. I suspect that the proportion is not large, and that it would drop precipitously if the APA tried to exercise any authority over its membership. To be quite frank, I have difficulty *imagining* the APA exercising authority; I cannot conceive what form such exercise might take. The bare possibility of such exercise, prescinding from questions of acceptability, would require a radical redistribution of educational power, a centralization of academic functioning that is at least unprecedented and probably unconstitutional. And how could the legitimate authority of an academic professional organization be reconciled with the professions's ruling value, academic freedom? Of course, the academic profession, especially the philosophical sector of it, may be a poor prototype for a profession.[4] But even the medical profession, which presents such a monolithic appearance to outsiders, claims only 50% membership in its own noisy and opinionated professional organization, the American Medical Association[5]; that proportion is, apparently, decreasing. And a complete account of the professions certainly ought to be able to handle the academic profession.

At present, then, we simply cannot address the professions as we address incorporated businesses, viz., as legal, if fictional, individuals. And as Richard DeGeorge has pointed out, the major area of *collective* responsibility that we now recognize, viz., corporate responsibility, is collective only from a point of view internal to the corporation: *we* hold a corporation responsible for a social task (e.g., supplying water in a certain district) *as an individual,* and it is by virtue of the assignment of *individual* responsibility to this fictive individual that we can speak intelligibly of the *collective* responsibility of the members of the firm to carry out the task.[6] It is this corporative model of responsibility assumption that seems to dominate the account of professionalization presented above, and that account is therefore presented with some problems by the comparative disorganization of the professions. The completion of professionalization on this model will require much more powerful and inclusive professional organizations; they would have to be capable of assuming liability for actions undertaken by their members, who would answer to their organizations alone for performances within the definition of professional practice. Should the development of such organizations be encouraged, at least for professions where no theoretical difficulties attend legitimization?

Several arguments against encouraging such organizations come immediately to mind. The object of professionalization is supposedly to serve the public interest, but the professionals of our experience seem to recognize many interests as legitimate, only some of which dovetail with the public good at any given time. Sets of professional goods that seem above to fit together neatly—personal profit, the cultivation of expertise, the protection of the individual client, and service to the public in a specialized capacity—do not always seem to work together in fact. Indeed, the most common popular judgment on the professions is that they tend to pursue one of the first three sets of goods to the exclusion of the others and of the fourth. Thus, the most common accusation leveled at professions is (*1*) that they are greedy, that the first priority of their professional practice is the resulting extrinsic reward of money, prestige, or power; their professional organizations simply function to protect their income potential. And if their pursuit of the dollar is cause for suspicion, their (*2*) pursuit of the aesthetic delights of expertise in their own professional art is simply infuriating. Nothing can raise a hospital patient's blood pressure faster than being treated as an "interesting case," an object of eager curiosity to hosts of interns and subject of fascinated experimentation for the attending physicians. Their enthusiasm over a magnificent opportunity to refine the art of medicine quite escapes the poor patient, who soon comes to resent being treated as irrelevant to the proceedings and less valuable than his disease. Will they keep him sick in order to continue studying him? And even when the professional's practice is very clearly oriented to the interests of the individual who has requested his professional services, (*3*) it is not clear that service to those interests is service to the society. We are reminded of the Mafia lawyers who energetically track down every thread of constitutional right that might help their unsavory clients, all for the purpose of defeating the public interest in putting criminals behind bars. Professionals seem to have commitments all over the place, in constant conflict with each other as well as with us. The unified ethical content suggested by their published ethics does not seem to be reflected in the world of practice nor in the official acts of the professional associations. It would appear that the protection of the public, then, requires that we continue to hold professionals individually accountable to that public (via lawsuit if necessary) and refrain from granting even more legal powers and immunities to the profession as a whole or to any organization within it, at least as presently structured. The sort of structural modifications that would be required by the public interest may be deduced from the discussion so far: each professional would have to be required to belong to his professional association, the association would have to be, like any corporation, prepared to assume financial liability for malpractice or misfeasance, and a network of professional ombudsmen would have to be maintained, to mediate lay complaints and collective responsibilities. And such modifications seem far in the future.

By way of summary of the present section to this point: against the neat and plausible account of professions and professionalization suggested at the outset, I have held that current attitudes and practices of and towards professionals fail to bear it out, and that the professions are far more disorganized, ethically and legally, than the attribution of collective responsibility and rights will permit.

One more failure should be mentioned as similarly significant in our investigation of this account of the professions. There are identifiable social needs, and there are professions serving them; but there does not appear to be any one-to-one correspondence between profession and need. To be sure, the ABA proclaims the legal profession "guardians of the Rule of Law,"[7] but the thrust of the legal profession seems to be much more toward the settlement of disputes, and it is from that need that lawyers are wont to trace their own origins. Guardianship of the law was traditionally assigned to statesmen, who despised the litigious sorts who represented clients in lawsuits; but the profession of "statesman" seems to have been lost in the development of the democracies. "Health care" also is a societal need, but the professional picture here is even more blurred. The profession of medicine is not the only profession interested in health care; it is not interested in all aspects of health care, and *is* interested in subjects (especially in medical research) that are only indirectly involved with protecting the health of the people of the society. The nursing profession, meanwhile, which is in constant, if subdued, conflict with the medical profession over jurisdictions of health care, regards its medically oriented work as continuous with its work in caring for people who have no health problems at all. The laboratory work that is directed to health care—from the biochemist's experimentation with inhibitors of enzymes to the technician's run of standard serum tests—is continuous with laboratory work dedicated to the development of more salable toothpaste. In short, the picture of professional health care is not one of a profession internally *dedicated* to the care of the people's health and assuming publicly the moral responsibility for that care. It is, rather, one of disparate preexistent professions *recruited* by the society to serve a perceived need, once the need was perceived, and only reluctantly working together on the assigned tasks. The same picture emerges in other professional areas: as the profession of "statesman" had disappeared, its expertise discredited in populist revolts, the profession of law was recruited to fill the breach; it has not done so very successfully.[8] More successful was the recruitment of the physicists in the second and third quarters of this century. The profession has been around, discussing the abstruse qualities of basic matter, for centuries; when military exigency suddenly required the development of highly sophisticated nuclear explosives and the rockets to deliver them, a whole profession moved easily into the task, putting together whole new industries, not to mention perfectly dreadful military capabilities, virtually overnight. Basic research now seems to be less useful to the government, and physics is retreating from a central position of social significance back to the academic place it once occupied. Nowhere along the line, in its brief period of national limelight, did the profession announce any professional dedication to public service. Rather, the profession had the expertise, knew where experts could be found and how to train more of them; and that expertise, whose existence predated the social need for it, was all that was required of them.

What we have, in short, in the relation between our professions and the state, is a highly fluid situation. Professions arise and disappear without public trust or social responsibility playing any particular part in their fate; they are recruited, singly or in groups, to assume a social responsibility more or less fitted to their talents, dismissed when the need is gone or a better solution to the problem is found, all without changing their nature as professions. How is it then, that professions exist, prior and posterior to social need, only partially and badly organized in professional associations, able somehow in this disorder to assume and carry on societal responsibilities when required to? A sketch of an answer to that question will constitute, in effect, another account of professionalization.[9]

This account of professionalization might begin, not with the notion of a "societal need," but with that of a "perennial activity." The inception of a profession would be found in the existence of a skilled activity—one in which expertise can be cultivated, if its practitioners desire to cultivate it—that some people will always take up, without reference to extrinsic motivations or rewards beyond the natural end of the activity. One class of perennial activities would be those enjoyable for their own sake. We are all familiar with this class, or we ought to be, for philosophical inquiry is one of them. Gambling is another, and the performing arts provide as many more examples. A second class of perennial activities would include those addressed to obvious and recurrent human need: cooking, or gardening, or healing the sick, or seeking (or preventing) the intervention of the deity to further human goods. In each of these activities, the art or skill may be cultivated for its own sake, i.e., for the sake of that simple and universal human satisfaction of doing something very well indeed. In most of them, practitioners will enjoy and profit from each other's company for purposes of exchanging information and practicing new techniques. In most of them, structured associations may (but need not) arise, to keep the like-minded practitioners in touch with each other and with the state of the art, eventually to formulate standards of excellence for the enterprise, perhaps to arrange conventions, contests, and prizes to encourage the attainment of higher levels of skill in practice, perhaps where relevant, to circulate literature of interest to the field. We have now "amateur" societies dedicated to philately, mycology, and organic gardening, not to mention scientology and theosophy, which differ from groupings of the learned professions only in that there is insufficient private or public demand for their skills to make any widespread practice profitable.

The beginnings of a profession might be here, then, in any skilled activity. The next logical step would be the one implicit in the last distinction above: to attempt to find a market for the skill, so that the practitioners would not be compelled to hold down jobs in addition to their chosen practice. In ordinary parlance, it is this step into "full-time," paid, status that makes the "professional"; Stephen Toulmin makes it his first criterion of professionalization that "Professionals engage in their chosen activities 'for a living' rather than 'for fun'—as amateurs or dilettantes."[10] Nothing about a commitment to public service is entailed here. If a charitable organization, for instance, decides that it needs much more money than it is getting by having its own members pass the hat, it may decide to hire a "professional" fund raiser, one who will charge the organization for his services, as an investment. In this case, "professional" means specifically "for hire," "personal-gain-oriented," in contrast to a pure public service orientation. For some professions, moreover, the orientation to the sale of professional services seems to be the defining mark of "membership" in the profession. In the case of the oldest of all the professions, for example, the activity that characterizes professional performance is certainly enjoyable in itself, but could never serve as a basis for group identity when carried on in an amateur status. It is specifically the decision to engage in the activity "full time," for pay, that creates the orientation toward the development of excellence in the art.

A profession crystallized, primarily or otherwise, around the common objective of personal gain might also find it worthwhile to create a professional organization to further the ends of the profession. Of all the varieties of association in our society, as a matter of fact, the most common seems to be of persons who earn their living in the same way, the labor unions being the prime example. Unions (and their relatives with other names) can keep track of markets and help the members sell their services at the highest possible price. They can ensure that members who have halted work temporarily in order to concentrate on efforts to secure higher remuneration will not be replaced, and their protest rendered useless, by similarly trained professionals. With the large numbers of an entire profession to back them up, unions can engage in political lobbying and public campaigns to secure legislation favorable to the profession, and so forth. But note that here again, the profession is not required to organize to establish its identity.

We have, then, two candidates for the essence of "professionalism," two possible *raisons d'être* for a profession: it can exist for the sake of excellence in the practice, or it can exist for the sake of profit for the practitioners. A third possible motivating principle overlaps with the first: the point of the activity might be the direct service rendered to identifiable others. The qualifiers "direct" and "identifiable" rule out professions in, e.g., scientific research, which may very well obtain information useful to someone somewhere in the long run, but whose criteria for excellence do not include reference to that use. A service orientation in a profession would originate in the clear and urgent need of another human being, to which we respond, in compassion, by offering help. Perhaps the best examples of such help should not be taken from the overworked and complicated files of the medical profession, but from those of much more recent health care and other "helping" professions: the geriatric nurses who staff nursing homes and homes for the elderly, the specialized pediatric staff who administer the schools for retarded and handicapped children, the funeral directors, social workers, family counselors, career advisors, and marriage consultants. What all these professions have in common is that until relatively recently, the helping function they perform was carried on completely by amateurs, perhaps slightly assisted by religious functionaries (whose duties, as originally understood, had very little to do with "pastoral" work, but emphasized almost exclusively the work of mediation between God and man). As the society becomes larger, more urbanized, specialized, and anonymous, not to mention richer, interpersonal functions increasingly migrate out of family and amateur purview into the realm of newly specialized expertise.

A peculiar feature of this type of professional practice, the feature that sets it off from the others and gives it moral direction, is the unique psychological and moral bond that normally forms between the professional and the individuals within his professional charge. Both aspects of this peculiar bond arise from the fact that in the situation of practice, the practitioner-client encounter, the professional is by definition, relative to the client, healthy, strong, knowledgeable, confident, and in control of the situation; the client is sick, weak, ignorant, frightened, and helpless. Prate as we will about the "autonomy" of the client, the fact remains that the client, at least at the outset of the relationship, must be passive and trusting, while the professional is the trusted agent. One moral bond, beyond the ordinary one obtaining in human relationships, is immediately generated: the client is terribly vulnerable, the professional has it easily within his or her power to take advantage of the client's helplessness for personal profit, and therefore the professional is under obligation not to do that. Beyond this prohibition, it may be, as is sometimes claimed, that positive responsibility is generated where the serious need for attention encounters the capacity to meet that need. I am not sure such a positive obligation could be established. If it can be, it is worth noting that this imperative, like the prohibition of harm, is generated merely by the inequality of the positions of professional and client, and by the situation of initial consultation, in which the individual approaches the professional for advice on *whether to assume* the role of "client"—prior to any contract, or to the establishment of any "professional-client" relationship that might be governed by law or professional ethics.

The psychological aspect of this bond, when present, stems from what Hume would call the natural sympathy among human beings. Seeing the need, and the helplessness, of another, and possessing the means to help, we are immediately moved to provide it. A more common way of describing the same phenomenon would be to say that the need, helplessness, and trust in us of another human arouses our parental feelings ("maternal instinct") and that we are naturally moved to protect and nurture that human. The generic name of this impulse is "love"; and so we can say that teachers, quite simply and innocently, come to love their students, physicians their patients, priests their flock, and nurses the persons in their charge. This curious and profound relationship gives a distinctive direction to professions where one-to-one practice is appropriate. From the moral aspect emerges the professional's obligation to protect the client from the dangers of professional performance at least, and from other dangers in his professional competence to combat; from the psychological aspect emerges a likelihood that the professional will in fact act protectively if a client is threatened by anything, including the larger society; and from the inarticulate combination of the two emerges the view that the professional has the right and duty to set aside the demands of any party, including the larger society, in the interests of protecting his client.

As these professions derive, primarily or otherwise, from the need of those they serve, their clients, the emphasis of a professional organization might well reflect that derivation in a high degree of "client-centered" concern. We would expect norms of professional conduct to be frankly based on the success of such conduct in providing help for individuals; we would expect these norms to encourage, in the professional, attitudes of protectiveness and support for clients, and to evaluate technical developments in the field by their immediate practical usefulness in helping clients. We would expect the professional organization to engage in political activities on behalf of client groupings, not groupings of professionals. Expecially where a newly professionalized service is carried on for pay, we would look for such professional organizations to set up stern criteria for membership, to exercise vigilance against "quacks" and other false practitioners seeking to victimize clients for profit, and to encourage legislation regulating the industry in which the profession does its work. Again, although such an organization would be highly useful in fulfilling professional tasks, it is not necessary; the help can be administered all the same without it.

On this account of professionalization, then, there are at least three possible routes to professional status prior to direct contact with the larger society: engagement in a perennial activity, development of a service for which a market exists, and provision of help to other individuals. Minimally, any profession, to be called by that name, must exhibit some degree of the first two characteristics: there must be an identifiable skilled activity that constitutes professional practice (including, as a necessary condition, some means of transmitting the "state of the art" to new generations of practioners,[12]) and the practice must be carried on for pay, "full time." Beyond that minimum, variety is the rule; a profession may exhibit only one professional form, pursue only one of the three possible goods; it may pursue one primarily and include the others as subsidiary goals; or it may, like the medical profession in its more exciting moments, exhibit all three pursuits in apparently equal proportion and noticeable conflict. Similarly, professional organizations, where they exist, may attempt, like the American Medical Association, to speak for all the goals of the profession, and so reflect, in their own internal conflicts and changes of direction, the failure of consensus in the profession as a whole. Or a profession may maintain, quite unofficially but quite effectively, a multiplicity of associations for its multiple pursuits. My own campus is not organized, but I have colleagues elsewhere who belong simultaneously to the American Philosophical Association (through which they advance the art of philosophy), the American Association of University Professors (to keep an eye on the teaching situation, with regard to academic freedom, "teacher evaluation" techniques, and so on), and a local affiliate of the AFL-CIO (to show up at contract time and protect their jobs and incomes.) The functions are not clearly divided among the associations even when all are simultaneously available, you will note; in this field, *nothing* is clearly divided.

Out of this conflict-filled background, how does it happen that professions can take on roles of tremendous social responsibility? For in the three possible commitments so far, the public good does not find a place; and even if a commitment to the public could be generated, the presence of the other three in some proportion would make perfect adherence to the commitment highly unlikely. Toulmin suggests,[13] and the suggestion is certainly plausible, that the profession's commitment to render service to the public is best understood as one major term of a "contract" between the society and the profession. Perceiving a social need, and the profession's competence to handle it, the society negotiates a deal with the profession: the society will confer the benefits and privileges of legal monopoly upon the group in return for a promise of public service, i.e., a promise to carry on professional practice in accordance with high standards of performance, for the public good. (Toulmin assumes that the deal is actually made between a legislative body and a professional guild or organization. This need not be the case; the legislature may simply announce that henceforth, all individuals who would call themselves "professionals" of a certain type must announce their willingness to adhere to a publicly promulgated code of ethics embodying the obligation to public service, in return for which the law will protect them in a quasimonopoly situation.[14]) To hold up its end of the deal, the profession incorporates its terms into its ethical fabric in some public way, typically issuing a professional ethic in which its preeminent willingness and sole competence to serve the people is set forth with maximum eloquence. This is why the codes emphasize public service to such a great extent: public service is the last professional commitment, not the first; it is the most doubtful of the profession's commitments, the one questioned most by its own members as well as by the society at large.

Now we have come full circle: the link between the profession and the larger society is recent, questionable, and inessential to the existence of either, but is of great benefit to the profession and eminently worth keeping. The writing of an eloquent and public-spirited "professional ethic" upon the formation of the link serves to explain and justify, to the general public, the conferral of privileges on the profession, even as it serves to reassure the legislature of professional trustworthiness, and convey the terms of the contract to the present and future members of the profession. The profession presents its ethic as a moral justification of its monopoly, and the public can justifiably hold a profession to that ethic to ensure that its needs will be met.

With this account of the professions, however—a variant of "social contract" theory, if we may refer to the original account as a "social generation" theory—the relationship between profession and society is shown to be a dynamic one, and additional options available for the profession and for the public some into focus. For instance, it becomes clear that a profession can refuse a public contract, if it finds any terms of public service unacceptable in light of its other commitments; various of the fine and performing arts have, on occasion, discussed the advisability of such refusal. For another instance, it becomes clear that the society can rearrange its contracts with the professions, to ensure a closer fit between professional directions and social need, without incurring the need for radical restructuring in those professions. A suggestion could be made, in illustration of this point, that the medical profession simply be dismissed as caretaker of public health, to be replaced by the nursing profession (or alternatively, by a separate group of "public health professionals"). Should that suggestion be implemented, medicine would continue to be practiced as before, as far as the treatment of individual patients is concerned; but authorization for treatment, third-party payments, hospital admissions, and so on would all be routed through the new social contractor. On the "social generation" account, such rearrangement would have to be called "deprofessionalization," which sounds much more drastic than the minimal adjustments that would actually be entailed by such a change. One virtue of the "social contract" account, then, is that it calls attention to the flexibility of the undertakings we have with and as professionals, and the range of possible relationships among which the public and the profession may choose.

Another virtue of the account, I will say by way of conclusion, is the legitimacy it provides for the preservation of a plurality of logically independent and potentially conflicting goods. The structure of a profession in fact permits and encourages conflicting pursuits, as noted; and this is its major strength, for the conflict reflects the genuine irreducibility of value orientations. It is dubious, as Thomas Nagel says, "that all value rests on a single foundation or can be combined into a unified system, because different types of values represent the development and articulation of different points of view, all of which combine to produce decisions."[15] The various possible conceptual origins of professions that I have tried to articulate in the foregoing are the developmental origins of such contrasting "points of view"; priorities must be assigned among these perspectives for individual decisions in conflict situations, but as general principles they retain their independent validity.

The value of the professions, then, for any advanced society, is not only that they perform social services on occasion, or assume responsibility for an area of social concern. It is also that, whether or not currently engaged in social service, they retain their nature as articulators of values. Even more, that in articulating and protecting potentially conflicting values, they truthfully reflect a moral universe that does not seem to be entirely coherent and that finds its best institutional expression in the permanent (and quarrelsome) coexistence of a plurality of institutional goods.

NOTES AND REFERENCES

[1]This paper is a revised version of a paper read at the inaugural meeting of the Society for Philosophy and the Professions, December 27, 1978, in conjunction with the annual meeting of the Eastern Division of the American Philosophical Association.

[2]See the account along these lines in Kenneth Kipnis' article in this volume, "Professional Responsibility and the Responsibility of Professions."

[3]Kurt Baier, "Guilt and Responsibility" in *Individual and Collective Responsibility,* P. French, ed. (Cambridge, Ma.: Schenkman Pub. Co., 1972), pp. 35–61.

[4]And perhaps it should avoid the whole process. See Alison Jaggar, "Philosophy as a Profession" *Metaphilosophy* 6: (1975), 100–116.

[5]That membership figure on the AMA is from Tristram Engelhardt's paper on "Physician's Responsibility and the Community of Physicians," read at a conference on Collective Responsibility in the Professions, Dayton, Ohio, October 27, 1978.

[6]Richard T. DeGeorge, "Moral Responsibility and the Corporation," read at the conference on Collective Responsibility in the Professions (n. 5).

[7]Cited by Kipnis, *op. cit.* p. 5.

[8]As Monroe Freedman points out in *Lawyers' Ethics in an Adversary System* (Indianapolis and New York: Bobbs-Merrill, 1975), portions of the professional ethic of law may be flatly incompatible with pursuit of the public interest, not just potentially but actually and always. See Stephen Toulmin, "The Meaning of Professionalism: Doctors' Ethics and Biomedical Science" in *Knowledge, Value and Belief,* H. T. Engelhardt Jr. and D. Callahan, eds., (Hastings-on-Hudson, New York: Hastings Center, 1977), 264–268.

[9]Some elements of the following account are borrowed from Engelhardt, *op. cit.,* n. 5.

[10]Toulmin, *op. cit.,* p. 256.

[11]Kipnis, *op. cit.* p.11, n. 6.

[12]See Toulmin, *op. cit.*.

[13]*Op. cit.* pp. 256, 258.

[14]See Engelhardt, *op. cit., n. 5.*

[15]"The Fragmentation of Value," in *Knowledge, Value and Belief, op. cit.,* p. 290.

Moral Responsibility in Professional Ethics[1]

Gerald J. Postema

Professionals generally acknowledge gravely that they shoulder special responsibilities, and believe that they should conform to "higher" ethical standards than laypersons.[2] Yet, doctors, lawyers, engineers, and indeed all other types of professionals also claim special warrant for engaging in some activities that, were they performed by others, would be likely to draw moral censure.[3] Skeptical of this claim to special license, Macaulay asked about lawyers (and most of my examples in this essay shall be drawn from law since that's where my experience lies), "[w]hether it be right that a man should, with a wig on his head, and a band round his neck, do for a guinea what, without these appendages, he would think it wicked and infamous to do for an empire."[4] This conflict may trouble the layperson, but for the professional who must come to grips with his or her professional responsibilities it is especially problematic.

Montaigne offered one solution, the complete separation of personal and professional lives. "There's no reason why a lawyer . . . should not recognize the knavery that is part of his vocation," he insisted. "An honest man is not responsible for the vices or the stupidity of his calling."[5] The key to maintaining both professional and personal integrity in the face of professionally required "knavery" was, Montaigne thought, scrupulously to keep the two personalities apart: "I have been able to concern myself with public affairs without moving the length of my nail from myself . . . . The mayor and Montaigne have always been two people, clearly separated."[6]

Montaigne's solution is tempting. Maintaining a hermetically sealed professional personality promises to minimize internal conflicts, to shift responsibility for professional "knavery" to broader institutional shoulders, and to enable a person to act consistently within each role he or she assumes. But for this strategy to succeed, the underlying values and concerns of important professional roles, and the characteristic activities they require, must themselves be easily segregated and compartmentalized. However, since there is good reason to doubt they can be easily segregated, Montaigne's strategy risks a dangerous simplification of moral reality. Furthermore, in compartmentalizing moral responses one blocks the cross-fertilization of moral experience necessary for personal and professional growth. This essay considers whether it is possible to follow Montaigne's suggestion and to separate one's private and professional personalities without jeopardizing one's ability to engage in professional activities in a morally and professionally responsible way. The central issue I address is not whether there is sufficient justification for a distinct professional code for professionals, for lawyers in particular, but whether, given the need for such a code, it is possible to preserve one's sense of responsibility. I argue that such preservation is not possible when a professional must adopt Montaigne's strategy in order to function well in the professional role. I contend that a sense of responsibility and sound practical judgment depend not only on the quality of one's professional training, but also on one's ability to draw on the resources of a broader moral experience. This, in turn, requires that one seek to achieve a fully integrated moral personality. Because this is not possible under the present conception of the professional's role, as perhaps is well-exemplified by the American Bar Association's Code of Professional Responsibility, that conception must be abandoned, to be replaced by a conception that better allows the professional to bring his or her full moral sensibilities to play in his or her professional role.

I. Moral Distance and the Perspective of the Responsible Person

It is not uncommon for professionals to face dilemmas caused by the clash of important principles implicit within the professional code. In law, a good example of this is the problem posed for a criminal defense lawyer by a client who announces a firm intention to commit perjury at trial.[7] Here, the deeply embedded principle of confidentiality[8] conflicts sharply with the equally important duty of candor before the court.[9]

But this is not the sort of clash Montaigne had in mind. Indeed, similar moral quandaries and conflicts are common outside of professional contexts. Rather, Montaigne draws attention to the conflict between principles of professional ethics and concerns of private morality. The requirements of professional ethics can sometimes move some distance from the concerns of private or ordinary morality, a phenomenon we might call *moral distance*. The range of practical considerations that alone are relevant to a proper ethical decision in a professional role is the *moral universe* of that role.[10] For many professional roles the moral universe of the role is considerably narrower than that of ordinary morality, and, when the two overlap, they often assign different weights to the same set of considerations. This often gives rise to conflicts, as the following cases illustrate.

The first example involves the duty of the criminal defense attorney to maintain client confidentiality. In the course of preparing a defense for a criminal trial in Lake Pleasant, New York, the client told his attorneys that he was responsible for three other murders unrelated to the pending case. The lawyers visited the location where one of the bodies had been hidden and confirmed the client's story. Nevertheless, they maintained silence for six months and refrained from disclosing the whereabouts of the body to the authorities or to the family of one of the victims, which had sought their help in locating the missing victim.[11] The duty of confidentiality, which here protects the client against self-incrimination, clearly forbade disclosure in this case,[12] even though the attorneys' personal inclinations were to disclose.

The second example, illustrated by the case of *Zabella v. Pakel,*[13] concerns the lawyer's use of legally available defenses to circumvent enforcement of the client's moral obligations. At the time of suit, defendant Pakel was president and manager of the Chicago Savings and Loan Association. In earlier and less fortunate circumstances, he had borrowed heavily from his employee, Zabella. Pakel gave a note for the borrowed sums, but, before the debt was paid, he declared bankruptcy. Zabella sued Pakel in 1954, contending that the defendant had made a subsequent promise to pay the outstanding debt. Under Illinois law, a new promise is sufficient to block the defense of discharge in bankruptcy, but because Pakel's promise was not in writing the defendant was able successfully to assert the statute of limitations as a defense. Despite the moral obligation of the affluent Pakel to repay the old debt, the statute of limitations blocked legal enforcement.[14] Many lawyers would argue that for Pakel's lawyer to have failed to assert the statute of limitations defense would have been a gross violation of professional responsibility.[15] Yet from the point of view of ordinary morality, the lawyer was implicated in the moral wrong done by the client. Professionally upright activities advancing the client's morally questionable, though legally sound, schemes are paradigmatic of the knavery to which Montaigne referred.

If we admit that some distance is likely to separate private and public morality, then several additional questions come to mind. How are these moralities related? How are conflicts between them to be resolved? Do they share a common foundation that could provide elements for resolution of the conflicts? One might seek a casuistry for a broad range of dilemmas that are likely to arise in any profession (which could then perhaps be taught to those preparing to enter the profession). Alternatively, one might seek some general account of the relationship between the principles governing each morality from which one could derive solutions to particular problems as they arise. The present approach to legal ethics, as embodied in the Code, largely utilizes the first strategy; philosophers tend to be partial to the second. Both can be useful, yet they both represent an approach to the problems posed by moral distance that is inadequate in important ways.

The first problem with these strategies is that they are only of limited usefulness in practice. Casuistry gives us solutions to isolated problems, but no strategy for resolving new problems. The systematic strategy is also not likely to help us in situations in which we experience the most puzzlement, since the moral dilemmas facing a professional generally cannot be reduced to a single perspective. Thus, a lawyer's personal and professional concerns do not have the collective uniformity necessary for the construction of a general scheme of principles and priority rules. On the contrary, these concerns are characterized by a complexity and a variety that resist reduction to a uniform scale.[16] As Thomas Nagel has argued, we are subject to moral and other motivational claims of very different kinds, because we are able to view the world from a variety of perspectives, each presenting a different set of claims. Nagel maintains that conflicts between perspectives

> [c]annot . . . be resolved by subsuming either of the points of view under the other, or both under a third. Nor can we simply abandon any of them. There is no reason why we should. The capacity to view the world simultaneously from the point of view of one's relations to others, from the point of view of one's life extended through time, from the point of everyone at once, and finally from the detached viewpoint often described as the view *sub specie aeternitatis* is one of the marks of humanity. This complex capacity is an obstacle to simplification.[17]

Are we left without any rational means for resolving this conflict? Nagel rightly resists this skeptical response.[18] The conflict Nagel describes shows only that it may be futile to search for a general reductive method or a clear set of priority rules to structure our basic concerns. There is always likely to be a significant gap between general practical theory and actual decision and practice.

In Aristotle's view, this gap is bridged by the faculty of *practical judgment*[19]—what he called practical wisdom.[20] Our ability to resolve conflicts on a rational basis often outstrips our ability to enunciate general principles. In doing so, we exercise judgment. Judgment is neither a matter of simply applying general rules to particular cases nor a matter of mere intuition. It is a complex faculty, difficult to characterize, in which general principles or values and the particularities of the case both play important roles. The principles or values provide a framework within which to work and a target at which to aim. But they do not determine decisions. Instead, we rely on our judgment to achieve a coherence among the conflicting values that is sensitive to the particular circumstances. Judgment thus involves the ability to take a comprehensive view of the values and concerns at stake, based on one's experience and knowledge of the world. And this involves awareness of the full range of shared experience, beliefs, relations, and expectations within which these values and concerns have significance.

In professional contexts there is much need for practical judgment in this Aristotelian sense. Judgment, however, is both a disposition—a trait of character—and a skill that must be learned and continually exercised. It is important, then, if we are seriously to consider matters of moral responsibility in professional contexts, that we pay attention to the conditions of development of this disposition and the exercise of this skill.

The second difficulty with the current approach to questions raised by the conflict between private and professional moralities is that it rests on a mistaken view of moral judgment and moral experience. Practical moral reasoning is wrongly viewed as strictly analogous to theoretical reasoning, the central objective of which is to arrive at correct answers to specific problems. This view of moral reasoning and experience is too narrow, for moral reasoning is not so singularly outcome-determinative. Our evaluations of ourselves and our actions depend not only on getting our moral sums right, but also on having the appropriate attitudes and reactions to the moral situation in which we act. Let me illustrate.[21]

Consider the truck driver who, through no personal fault, hits and seriously injures a child. It may be correct to say that, since the trucker drove with care and could not have avoided hitting the child, the driver is guilty of no wrong and thus is not blameworthy. However, consider the accident from the driver's point of view as someone personally involved in it. There is a very important difference between the driver's likely reaction and that of an uninvolved spectator. Both may feel and express regret, but the nature and behavioral expression of this regret will be quite different. The driver's direct, personal (albeit unintentional) involvement in the accident alters the structure of the moral situation and the driver's attitude toward it. The difference in the emotional responses of the driver and the spectator will be reflected in the way these feelings are expressed. The driver may attempt to make restitution in the hope that the injury caused can be repaired. The spectator may offer help, or contribute money for hospital bills, or even visit the child, but these actions would be understood (by the spectator and by us) as expressions of pity, kindly concern, or perhaps generosity. From the driver, these same actions would be intended and understood as expressions of a special form of regret. Suppose, however, that the driver takes the attitude of the uninvolved spectator, perhaps expressing detached regret, but feeling no need to make restitution. The driver can rightly argue that he or she was not to be blamed for the accident, that no personal wrong had been done. In doing so, the driver could perhaps be rightly said to have gotten the moral sums right. But in asserting this defense quite sincerely and too quickly, the driver would also reveal a defect of character—a defect much deeper and more serious than a lack of generosity. Morality seems to require not only that one be able to apply moral principles properly to one's own or another's conduct, but also that one be able to appreciate the moral costs of one's actions, perhaps even when those actions are unintentional. By "moral costs" I mean those features of one's action and its consequences touching on important concerns, interests, and needs of others that, in the absence of special justification, would provide substantial if not conclusive moral reasons against performing it.[22]

Similarly, in cases in which obligations to other persons are correctly judged to be overridden by weightier moral duties, with the result that some injury is done, it is not enough for one to work out the correct course of action and pursue it. It is also important that one appreciate the moral costs of that course of action. This appreciation will be expressed in a genuine reluctance to bring about the injury, and a sense of the accompanying loss or sacrifice. It may even call for concrete acts of reparation: explaining and attempting to justify the action to the person injured, or making up the loss or injury to some extent. This is one way in which the moral status of the principle or right that was violated is acknowledged, and the moral relations between the parties affirmed and, when necessary, repaired.

Moral sentiments are an essential part of the moral life. The guilt or remorse one feels after mistreating a person is not merely a personal sanction one imposes on oneself after judging the action to have been wrong; it is the natural and most appropriate expression of this judgment. Similarly, the outrage we feel at injustice done to another and the resentment we feel at wrong done to ourselves are not just the emotional coloring of detached moral judgments, but the way in which we experience and express these judgments. Thus, morality is not merely a matter of getting things right—as in solving a puzzle or learning to speak grammatically—but a matter of relating to people in a special and specifically human way.

It must be admitted that these elements of practical wisdom and moral sentiment are not needed to understand the proper performance of duties in *some* professional or occupational roles. We need spend little time worrying about whether our understanding of the duties of bank clerks or auto mechanics properly allows for these elements. A person's moral faculties are not extensively engaged in the characteristic activities of these roles. In contrast, the charateristic activities of, say, a lawyer's role demand a much greater involvement of the moral faculties.

For these reasons, we must approach the problems of professional ethics from a perspective that recognizes the importance of practical judgment and moral sentiment. The notion of professional responsibility should take on a different and broader meaning. The primary concern is not with the definition, structuring, and delimitation of a lawyer's professional *responsibilities* (his official concerns and duties), nor with those situations in which the lawyer is to be held professionally *responsible (i.e.,* liable to blame or sanction). Rather, the concern is with responsibility as a virtue or trait of character.[23]

The focus, then, is on the notion of a *responsible* person—or perhaps better, on the notion of a person's *sense of responsibility*. My concern in the rest of this essay is to explore the ways in which institutional structures and public expectations, as well as the personal attitudes and self-conceptions of professionals, affect the development of this sense of responsibility.

II. Moral Distance and the Call for "Deprofessionalization"

Before we proceed with the argument, however, we must consider briefly a more radical solution to the problems posed by the phenomenon of moral distance. It is sometimes argued that the dissonance between the dictates of professional and private morality is *itself* a symptom of a deep social and moral problem. The solution, it claimed, lies in a "deprofessionalization" of professional roles[24] that would reduce all professional responsibilities to species of private morality. This view holds that it either is or should be the case that the duties and responsibilites of a professional are no different from those of any lay person facing a similar moral problem.

This approach to professional responsibility makes two serious mistakes. First, it rests on a mistaken objection to what we might call the "exclusionary character" of professional morality.[25] We have already seen that the moral universe of a professional role characteristically is narrower than that of ordinary morality. But since the moral universe defines the range of considerations that a role agent may take into account in choosing a course of action, it is possible that otherwise relevant considerations may be effectively excluded from the agent's deliberation. Thus, cases may arise in which an agent is required by his role to act without considering the full range of moral reasons before him; rather, he must consider only those moral reasons within his particular moral universe. Critics argue that this is both irrational and morally suspect since it denies the role agent his essential rational autonomy.

But it is not difficult to show that there is nothing inherently irrational or morally objectionable in this exclusionary character of professional morality. Some examples should make this clear. Suppose I have a tendency to spend my paycheck frivolously, with the result that my monthly bills pile up unpaid. To avoid this dangerous situation I adopt the policy of paying bills first and spending on pleasure only what is left over. I know that in the absence of this personal policy, I am liable to be moved by immediate desires to postpone paying my regular bills. Suppose that after adopting this policy I come to believe, at the end of a very tiring month, that I deserve a weekend holiday, although I know I can afford it only if I postpone payment on several bills. I decide, however, by appealing to my policy, to pay the bills. In this case, the policy operates not as an additional factor to consider along with the good the holiday would do for me and the difficulty I will face if my bills are not paid, but as an exclusionary factor. Indeed, it may be true that this month no particular harm would come from my postponing payment of the bills. The policy, however, instructs me not to consider other factors; it excludes them from consideration and provides a reason for not acting on the balance of reasons in this case. Of course, the exclusionary policy must itself be supported by sufficient reasons, but they need not apply directly to the particular case. Similarly, the moral appropriateness of exclusionary reasons is evident in any standard case of promising. Suppose I promise to drive you across town to your doctor's appointment tomorrow. In this case, considerations of cost and inconvenience that otherwise might be sufficient to persuade me not to take the trip are excluded—the exclusion is not absolute. In neither of these cases, then, am I subject to a charge of irrationality or moral irresponsibility.

Second, the call for deprofessionalization fails to appreciate the important social value of professional roles having this exclusionary character. We design social institutions to perform important tasks and to meet social needs or serve important social values. To carry out these tasks we design specific roles within the institutional framework and entrust them with responsibility over a particular range of social concerns. The domain of practical concerns determined by the basic tasks of the role is the moral universe of that role. Thus, social and professional roles represent an important division of social and moral labor. And carefully defined boundaries of concern and responsibility are needed for the efficient and successful achievement of important social goals served by the division of labor.

Thus, there is nothing objectionable in general, nor anything unique, about the phenomenon of distance between private and public morality. Critical attention must be turned, rather, to the way in which the moral universe of a given role is defined and structured, and the effect this has on the professional's ability to act responsibly in the moral universe so defined.

III. Responsible Action Under the Standard Conception

The central problem I am concerned with is whether, given the fact of moral distance, it is possible to retain and act out of a mature sense of responsibility in a professional role. In this section, I argue that because of particular social and psychological features of professional roles, the pressures and tensions of acting and deliberating within such roles pose a serious threat to responsible professional behaviour. In addition, I hope to show that the atrophy of the professional's sense of general moral responsibility is a serious and costly matter. If this argument is correct, we have in general discovered an important dimension along which to evaluate competing conceptions of professional roles, and in particular we have a strong reason for radically rethinking the standard conception of the lawyer's role.[26] This standard conception[27] is marked by two central ideas:

(i) Partisanship: the lawyer's sole allegiance is to the client; the lawyer is the partisan of the client. *Within,* but all the way *up to,* the limits of the law, the lawyer is committed to the aggressive and single-minded pursuit of the client's objectives.

(ii) *Neutrality:* once he has accepted the client's case, the lawyer must represent the client, or pursue the client's objectives, regardless of the lawyer's opinion of the client's character and reputation, and the moral merits of the client's objectives. On this conception, the lawyer need not consider, nor may he or she be held responsible for, the consequences of professional activities as long as they remain within the law and act in pursuit of the client's legitimate aims.[28] Thus, the proper range of the lawyer's concern—the boundaries of the lawyer's "moral universe"—is defined by two parameters: the law and the client's interests and objectives. These factors are the exclusive points of reference for professional deliberation and practical judgment. I contend that, far from encouraging the development and preservation of a mature sense of responsibility, the standard conception tends seriously to undermine it. To show why this is so I must sketch briefly what might be called "the problem of responsibility."

The problem is suggested in a rather grand way by Sartre in a familiar argument from his early existentialist period.[29] Sartre insisted that to take role moralities seriously is to fail to take responsibility for oneself and one's actions. The essential property of human consciousness, according to Sartre, is its absolute freedom—the capacity to define oneself in action independently of one's role or roles. Roles, however, come "ready-made," packaged by society. When acting in a role, one simply acts as others expect one to act. Simply to identify with one's role is to ignore the fact that one is free to choose not to act in this way. In Sartre's view, it is therefore essential that one be capable of walking away from one's role. Furthermore, although it is psychologically possible to identify deeply with one's role, doing so is, in Sartre's view, morally unthinkable and a form of bad faith. Identification is a strategy for evading one's freedom and, consequently, one's responsibility for who one is and what one does. By taking shelter in the role, the individual places the responsibility for all of his or her acts at the door of the institutional author of the role.[30]

Sartre's problem arises from the fact that in addition to moral distance, there is a second dimension—psychological distance—characteristic of the experience of persons who assume professional roles. Echoing Sartre, Goffman notes that

> in performing a role the individual must see to it that the impressions of him that are conveyed in the situation are compatible with role-appropriate personal qualities effectively imputed to him: a judge is supposed to be deliberate and sober; a pilot, in a cockpit, to be cool . . . . These personal qualities, effectively imputed and effectively claimed, combine with a position's title, where there is one, to provide a basis of *self-image* for the incumbent and a basis for the image [others will have of his role]. A self, then, virtually awaits the individual entering a position . . . . [31]

Psychological distance is especially characteristic of professional roles. As Goffman seeks to show in his essay, one can identify with, or distance oneself in varying degrees from, this available self-image. The more closely one identifies with one's role, the more one's sense of self is likely to be shaped by the defining features of the role.[32] At one extreme, maximal identification is characterized by an unquestioning acceptance of the duties and responsibilities of one's role. For the person who maximally identifies with his role, the response "because I am a lawyer," or more generally "because that's my job," suffices as a complete answer to the question "why do that?" One minimally identifies, on the other hand, when one conforms only to avoid the external consequences of failing to do so, in no way internalizing the role or its basic principles. Several possible intermediate states separate these extremes.[33]

Thus, in addition to the dimension of moral distance between private and professional morality there is the dimension of psychological distance between oneself, or one's moral personality, and one's role. Furthermore, these two dimensions are interrelated: the extent to which one identifies with one's role is a function not only of one's moral personality, but also of the moral distance between role morality and one's private morality. The opposite influence is also possible. Acting and deliberating within the special moral universe of any role that involves a large investment of one's moral faculties will tend to shape one's moral personality and, thus, one's inclination to identify with the role. The problem of responsibility lies in the fact that as the moral distance between private and professional moralities increases, the temptation to adopt one or the other extreme strategy of identification also increases; one either increasingly identifies with the role or seeks resolutely to detach oneself from it. Under either extreme, however, one's practical judgment and sense of responsibility are cut off from their sources in ordinary moral experience. Yeats warned that "once one makes a thing subject to reason, as distinguished from impulse, one plays with it, even if it is a very serious thing."[34] We might say, paraphrasing Yeats, that the artificial reason of professional morality, which rests on claims of specialized knowledge and specialized analytical technique, and which is removed from the rich resources of moral sentiment and shared moral experience in the community, tempts the professional to distort even the most serious of moral questions.

The problem of responsibility is especially troubling for the legal profession. The risk of severing professional judgment from its moral and psychological sources is particularly strong in a profession that serves a system of institutionalized justice.[35] As a result, the problem of developing a sense of personal responsibility is critical for the legal profession. First, the factors inducing maximal psychological identification are strong.[36] Publicly dedicated to serving socially valued institutions, the lawyer occupies a key role in society, enjoying considerable social status. The lawyer's claim to specialized knowledge and skill puts the lawyer in a position of power relative to the client. These facts, in addition to the important intrinsic satisfaction of exercising these special skills, encourage a high degree of role identification. As a result, principles of professional responsibility, originally justified on functional grounds, take on independent value and significance for lawyers. Professional integrity becomes a mark, often the most significant mark, of personal integrity.

Second, the characteristic activities of lawyering often require the lawyer to act in the place of the client, and thus require the direct involvement of the lawyer's moral faculties—*i.e.*, the capacities to deliberate, reason, argue, and act in the public arena. All professionals, and many persons in service-oriented occupations, do things for a client that the client is unable or unwilling personally to do. But, unlike the lawyer, the physician or auto mechanic acts only to provide services for the client. The lawyer also acts as the client's *agent*. Although an individual may employ a physician or mechanic to operate on one's body or automobile, the work of these professionals is in no sense attributable to the patient or customer. When the lawyer acts to secure the client's interests, however, the attorney often acts, speaks, and argues in the place of the client. The lawyer enters into relationships with others in the name of the client. When arguing in the client's behalf, the attorney often presents the client's arguments; when acting, the lawyer is often said to be "exercising the client's rights." And what the lawyer does is typically attributable to the client.[37] Thus, at the invitation of the client, the lawyer becomes an extension of the legal, and to an extent the moral, personality of the client.

Since the lawyer often acts as an extension of the legal and moral personality of the client, the lawyer is under great temptation to refuse to accept responsibility for his or her professional actions and their consequences. Moreover, except when these beliefs coincide with those of the client, the lawyer lives with a recurring dilemma: the need to engage in activities, make arguments, and present positions that he or she does not personally endorse or embrace. The lawyer's integrity is put into question by the mere exercise of the duties of the profession.[38]

To preserve personal integrity, the lawyer must carefully distance him or herself from these activities. Publicly, the attorney may sharply distinguish between statements or arguments made on behalf of the client and statements on which his or her professional honor is staked.[39] The danger in this strategy is that a curious two-stage distancing may result. First, the lawyer distances himself or herself from the argument: it is not one's own argument, but that of the client. The lawyer's job is to construct the arguments; the task of evaluating and believing them is left to others. Second, after becoming thus detached from the argument, the lawyer is increasingly tempted to identify with this stance of detachment. What first offers itself as a device for distancing oneself from personally unacceptable positions becomes a defining feature of one's professional self-concept. This, in turn, encourages an uncritical, uncommitted state of mind, or worse, a deep moral skepticism. When such detachment is defined as a professional ideal, as it is by the standard conception, the lawyer is even more apt to adopt these attitudes.

The foregoing tensions and pressures have sources deep in the nature of the lawyer's characteristic activities. To eradicate them entirely would be to eliminate much of what is distinctive and socially valuable in these activities. Nevertheless, these tensions can be eased, and the most destructive tendencies avoided, if lawyers have a framework within which they can obtain an integrated view of their activities both within the role and outside it. The framework must provide the resources for responsible resolution of the conflicts that inevitably arise. The standard conception of the lawyer's role, however, fails notably on this score. Clearly, the standard conception calls for a sharp separation of private and professional morality in which, to quote Bellow and Kettleson, "the lawyer is asked to do 'as a professional' what he or she would not do 'as a person.' "[40] The conception requires a public endorsement, as well as private adoption, of the extreme strategy of detachment. The good lawyer is one who is capable of drawing a tight circle around one's self and one's client, allowing no other considerations to interfere with one's zealous and scrupulously loyal pursuit of the client's objectives. The good lawyer leaves behind his or her own family, religious, political, and moral concerns, and becomes devoted entirely to the client.[41] But since professional integrity is often taken to be the most important mark of personal integrity, a very likely result is that a successful lawyer is one who can strictly identify with this professional strategy of detachment. That is, the standard conception both directly and indirectly *encourages* adoption of one or the other of the extreme strategies of identification. But, as we have seen, both strategies have in common the unwanted consequence that practical deliberation, judgment, and action *within* the role are effectively cut off from ordinary moral beliefs, attitudes, feelings, and relationships—resources on which responsible judgment and action depend. This consequence is very costly in both personal and social terms.

Consider first the personal costs the lawyer must pay to act in this detached manner. The maximal strategy yields a severe impoverishment of moral experience. The lawyer's moral experience is sharply constrained by the boundaries of the moral universe of the role. But the minimal strategy involves perhaps even higher personal costs. Since the characteristic activities of the lawyer require a large investment of his or her moral faculties, the lawyer must become reconciled to a kind of moral prostitution. In a large portion of the lawyer's daily experience, in which he or she is acting regularly in the moral arena, the lawyer is alienated from personal moral feelings and attitudes, and indeed from his or her moral personality as a whole. Moreover, in light of the strong pressures for role identification, it is not unlikely that the explicit and conscious adoption of the minimal identification strategy involves a substantial element of self-deception.

The social costs of cutting off professional deliberation and action from their sources in ordinary moral experience are even more troubling. First, cut off from sound moral judgment, the lawyer's ability to do the job well—to determine the applicable law and effectively advise his or her clients—is likely to be seriously affected.[42] Both positivist and natural law theorists agree that moral arguments have an important place in the determination of much of modern law.[43] But the lawyer who must detach professional judgment from personal moral judgment is deprived of the resources from which arguments regarding the client's legal rights and duties can be fashioned. In effect, the ideal of neutrality and detachment wars against its companion ideal of zealous pursuit of client interests.

Second, the lawyer's practical judgment, in the Aristotelian sense, is rendered ineffective and unreliable.[44] In section I, I argued that, because human values are diverse and complex, one is sometimes thrown back on the faculty of practical judgment to resolve moral dilemmas.[45] This is as true within the professional context as outside of it. To cut off professional decision-making from the values and concerns that structure the moral situation, thereby blocking appeal to a more comprehensive point of view from which to weigh the validity of role morality, is to risk undermining practical judgment entirely.[46]

Third, and most importantly, when professional action is estranged from ordinary moral experience, the lawyer's sensitivity to the moral costs in both ordinary and extraordinary situations tends to atrophy. The ideal of neutrality permits, indeed requires, that the lawyer regard his professional activities and their consequences from the point of view of the uninvolved spectator. One may abstractly regret that the injury is done, but this regret is analogous to the regret one feels as a spectator to the traffic accident mentioned in an earlier example[47]; one is in no way personally implicated. The responses likely from a mature sense of responsibility appear morally fastidious and unprofessional from the perspective of the present ABA Code. This has troubling consequences: without a proper appreciation of the moral costs of one's actions, one cannot make effective use of the faculty of practical judgment. In fact, a proper perspective of the moral costs of one's action has both intrinsic and instrumental value. The instrumental value lies in the added safeguard that important moral dilemmas will receive appropriate reflection. As Bernard Williams argued, "only those who are [by practice] reluctant or disinclined to do the morally disagreeable when it is really necessary have much chance of not doing it when it is not necessary . . . [A] habit of reluctance is an essential obstacle against the happy acceptance of the intolerable."[48]

But this appreciation is also important for its own sake. To experience sincere reluctance, to feel the need to make restitution, to seek the other's pardon—these simply are appropriate responses to the actual features of the moral situation. In this way, the status and integrity of important principles are maintained in compromising circumstances, and the moral relations between persons are respected or restored.

Finally, the moral detachment of the lawyer adversely affects the quality of the lawyer–client relationship. Unable to draw from the responses and relations of ordinary experience, the lawyer is capable of relating to the client only as a client. The attorney puts his or her moral faculties of reason, argument, and persuasion wholly at the service of the client, but simultaneously disengages his or her own moral personality. The lawyer views himself or herself not as a moral actor but as a legal technician. In addition, the lawyer is barred from recognizing the client's moral personality. The moral responsibilities of the client are simply of no interest. Thus, paradoxically, the combination of partisanship and neutrality jeopardizes client autonomy and mutual respect (two publicly stated objectives of the standard conception), and yields instead a curious kind of *impersonal relationship*.

It is especially striking, then, that Charles Fried, the most sophisticated defender of these central ideals of the standard conception, should describe the lawyer as a "special purpose" friend.[49] Indeed, it is the contrast between the standard conception of the lawyer–client relationship and the characteristics of a relationship between friends that, on reflection, is likely to make the deepest impression. The impersonalism and moral detachment characteristic of the lawyer's role under the standard conception are not found in relations between friends. Loyalty to one's friend does not call for disengagement of one's moral personality. When in nonprofessional contexts we enter special relationships and undertake special obligations that create duties of loyalty or special concern, these special considerations must nevertheless be integrated into a coherent picture of the moral life as a whole. Often we must view our moral world from more than one perspective simultaneously.[50] As Goffman points out, roles are often structured with the recognition that persons occupying the role fill other roles that are also important to them. Room is left for the agent to integrate his or her responsibilities from each role into a more or less coherent scheme encompassing the agent's entire moral life.[51]

But it is precisely this integrated conception of the moral personality that is unavailable to the professional who adopts either the minimal or the maximal identification strategy. Either the moral personality is entirely fragmented or compartmentalized, or it is shrunk to fit the moral universe defined by the role. Neither result is desirable.

IV. Toward an Alternative Conception: The Recourse Role

The unavoidable social costs of the standard conception of professional legal behavior argue strongly for a radical rethinking of the lawyer's role. One alternative is a "deprofessionalization" of legal practice so as to eliminate the distance between private and professional morality. Deprofessionalization, however, would involve a radical restructuring of the entire legal system, reducing the complexity of the law as it currently exists so that individuals could exercise their rights without the assistance of highly specialized legal technicians. But, setting aside obvious questions of feasibility, to discredit this proposal we need only recall that deprofessionalization ignores the significant social value in a division of moral and social labor produced by the variety of public and professional roles.

A second, more plausible alternative is to recognize the unavoidable discontinuities in the moral landscape and to bridge them with a unified conception of moral personality. Achieving any sort of bridge, however, requires that lawyers significantly alter the way they view their own activities. Each lawyer must have a conception of the role that allows him or her to serve the important functions of that role in the legal and political system while integrating a personal sense of moral responsibility into the role itself. Such a conception must improve upon the current one by allowing a broader scope for engaged moral judgment in day-to-day professional activities while encouraging a keener sense of personal responsibility for the consequences of these activities.[52]

The task of forging a concrete alternative conception is a formidable one. To begin, however, it may be useful to contrast two conceptions of social roles: the fixed role and the recourse role.[53] In a fixed role, the professional perceives the defining characteristics of the role—its basic rules, duties, and responsibilities—as entirely predetermined. The characteristics may be altered gradually through social evolution or more quickly by profession-wide regulatory legislation, but as far as the individual practitioner is concerned, the moral universe of that role is an objective fact, to be reckoned with, but not to be altered.[54] Sartre, proponents of the standard conception, and advocates of deprofessionalization all rest their positions on the assumption that the defining features of each role remain fixed. But this assumption fits only some social roles. A bank clerk, for example, must follow set routines; little judgment is required, and the clerk has no authority to set aside the rules under which he or she acts or alter these rules to fit new occasions.[55] This is not troubling because the sorts of situations one is likely to face in such a job lend themselves to routine treatment.

In contrast, in a recourse role, one's duties and responsibilities are not fixed, but may expand or contract depending on the institutional objectives the role is designed to serve.[56] The recourse role requires the agent not only to act according to what is perceived to be the explicit duties of the role in a narrow sense, but also to carry out those duties in keeping with the functional objectives of the role.[57] The agent can meet these requirements only if he or she possesses a comprehensive and integrated concept of his or her activities both within and outside the role. Role morality, then, within a recourse role is not properly served by maximal identification with one's role. Nor can the role agent minimally identify with the role so as to abandon or disengage all personal morality or a basic sense of responsibility. Indeed, responsible professional judgment will rely heavily on a sense of responsibility.

If we perceive the role of the lawyer in our legal system as a recourse role, a viable solution to the problem of responsibility may be available. A recourse role conception forces the lawyer to recognize that the exercise of role duties must fully engage one's rational and critical powers, and one's sense of moral responsibility as well. Although not intended to obliterate the moral distance between professional and private moralities, a recourse role conception bridges that gap by integrating to a significant degree the moral personality of the individual with the performance of role responsibilities. Most significantly, this conception prevents the lawyer from escaping responsibility by relying on his or her status as an agent of the client or an instrument of the system. The lawyer cannot consider him or herself simply as a legal technician, since the role essentially involves the exercise of an *engaged* moral judgment.

V. Objections and Replies

Two initial objections to the recourse role conception should be addressed: (1) that it is paternalistic,[58] and (2) that it is unnecessary, since the standard conception adequately allows for the exercise of individual moral judgment by permitting the attorney to decide whether to accept or withdraw from representation.[59]

Both these objections are mistaken. Paternalism involves interfering with the actions of others or, in subtler cases, making judgments for others, for their own good. Engaging a lawyer's moral judgment in the day-to-day practice of law in no way entails paternalism—and will involve paternalism only if the lawyer personally holds strongly paternalistic moral views. There is an important distinction between evaluating alternatives in terms of the client's long range good and making decisions without the client's consent (or against the client's will), on the one hand, and exercising one's moral judgment regarding those alternatives, on the other. The recourse role conception is not paternalistic even in the very broad sense that the lawyer is encouraged to make the client's moral decisions for him or her. The moral judgments that the lawyer makes are made on that lawyer's own behalf; the attorney does not make the client's decisions. Indeed, it is an advantage of my proposal, in contrast to the standard conception, that it enables both lawyer and client to recognize and respect the moral status of the other. The lawyer–client relationship should not be any more paternalistic than a relationship between friends.

As to the second objection, it may be admitted that the ABA Code of Professional Responsibility, like many other such codes, does provide some room for the individual lawyer to exercise moral judgment in the acceptance of and withdrawal from representation. But the scope allowed for such judgment is limited, and the motivation skewed. The Code mandates refusal to accept employment only when the client clearly intends to bring an action "merely for the purpose of harassing or maliciously injuring any person,"[60] or seeks to bring an action for which no reasonable legal argument can be made.[61]

The lawyer may refuse to accept employment when his or her personal feelings are sufficiently intense that they diminish the ability effectively to represent the client,[62] although the Code permits such refusal only in fairly compelling circumstances.[63] Once the lawyer has undertaken to represent the client, however, permissive withdrawal is allowed only in a few restricted circumstances,[64] none of which includes the conscience of the lawyer. Permissive withdrawal for what might be termed "moral reasons" is condoned only when the client insists upon presenting a claim or defense for which no reasonable legal justification can be advanced,[65] or when the client insists upon an illegal course of action.[66] Withdrawal for moral reasons is required only when the lawyer knows that the client is conducting the litigation solely to harass.[67]

Thus, in cases in which the Code permits the lawyer to refuse employment, the rationale seems to be that a lawyer who has scruples about the client's proposed legal action is not likely to be able adequately to serve that client.[68] And such scruples generally are not permitted at all once the attorney agrees to represent a client.[69] This point of view encourages lawyers to steel themselves against such scruples and to view them as strictly personal feelings that have no place in professional behavior—a kind of unbecoming moral squeamishness. It is hard to dismiss the thought that this reaction is at bottom morally cynical.

Charles Fried raises a more sophisticated objection to the conception of the lawyer's role developed above. On this point, he insists that the law must respect the autonomy and rights of individual citizens.

According to Fried, one way we respect individual rights is through the creation of specific rules and institutions to protect them. But when the legal system is so complex that the ordinary person is unable to exercise his or her rights without an expert advisor, the interjection of a lawyer's moral judgment might prevent the individual from exercising fully these lawful rights. In this way, "the law . . . would impose constraints on the lay citizen [implicitly] . . . which it is not entitled to impose explicitly."[70] Thus, Fried argues, although no rights are violated when a pornography venture fails because a person refuses on moral grounds to lend it funds, rights are violated when "through ignorance or misinformation about the law [because of an attorney's moral judgment] an individual refrains from pursuing a wholly lawful purpose."[71]

The problem with this argument is that while we might agree that individual autonomy and rights should be respected, we might still deny that it is the lawyer's moral as well as role duty to assist the client in any lawful exercise of a client's legal rights. Fried's mistake is to confuse *moral* rights with *legal* rights. He fails to distinguish the rights and sphere of autonomy defined by *morality* from the rights and area of free action defined by law. The area of legally permitted behavior need not coincide with that circumscribed as a matter of individual moral right. A lack of fit between legal and moral rights is most obvious when legal rights appear unjust or otherwise morally objectionable. But lack of fit may occur even when the legal system is morally ideal. For various reasons the law paints the canvas of legal relations, regulations, and rights with broad strokes. Many forms of social behavior otherwise harmful or morally objectionable are left unproscribed in light of the moral or social costs of enforcement.

Fried complains that if the services of the lawyer were restricted, the law would, in effect, implicitly impose constraints that it is not entitled to impose explicitly.[72] But this is not the case in the instances I am considering. For reasons of administrative efficiency, a particular restriction may not be imposed, but it does not follow that the lawmaker is not *entitled*—because it would violate someone's rights—to impose such a restriction.

Not all rights, powers, or permissions defined by the law protect moral rights. Fried's argument has force only for those legal rights that protect important individual moral rights and fundamental political liberties. It does not necessarily hold for other legal rights. My argument is not that the exercise of these legal rights is never justified. I merely contend that it is not always morally right for the individual to exercise these rights, and that it may, in particular instances, be wrong for a lawyer to help in their exercise. Thus, it is a matter of *moral argument* whether in particular cases it is appropriate for the lawyer to assist in what may be considered a morally questionable exercise of clear legal rights.

However, Fried seems to qualify his commitment to the standard conception of the lawyer's role. He rejects the contention, advanced by proponents of the standard conception, that whatever is legally permissible for a lawyer is morally permissible as well.[73] He has no trouble with a lawyer who assists his client in exercising his legal rights when only the collective good or abstract interests are adversely affected.[74] But Fried is troubled by those activities of a lawyer that cause harm to specific persons known to the lawyer and client, such as the adverse witness who may be abused by the defense lawyer. To resolve this tension, he distinguishes between personal wrongs (wrongs *personally* committed by the lawyer in the course of carrying out professional activities) and institutional wrongs (wrongs worked by the rules of the legal system).[75] However, in the absence of some independent principle by which to distinguish between personal and institutional wrongs, Fried seriously begs the very question at issue. The question can be rephrased: for what harms or wrongs done to specific individuals other than the client is the lawyer personally responsible, and what harms or wrongs must be laid at the door of the system itself?

Fried does suggest one answer, which turns on whether the description of the harmful action in question essentially refers to law or legal institutions. This principle can be illustrated by the difference between (a) abusing a witness in court and (b) exploiting unfairly the defense of the statute of limitations. In the second case, the harmful action is formally defined by the procedural rules of the legal system itself. Thus, legal institutions created the occasion for the action and are essential to the definition of the action itself. There is, for example, no action fitting the description "asserting the defense of the statute of limitations" outside a specific legal context, just as there is no action fitting the description "hitting a home run" outside the context of baseball. So, says Fried, the action is not personal, it is institutional—an action of the system, not of the individual lawyers.[76] In contrast, the act of abusing the witness is a personal act of the lawyer. Although it occurs in a legal context, it is an action that can be done within and outside that context. There is nothing essentially institutional about it. Thus, Fried concludes, "[t]he lawyer is not morally entitled . . . to engage his own person in doing personal harm to another, though he may exploit the system for his client even if the system consequently works injustice."[77]

Fried's distinction between personal and institutional wrongs is open to criticism on two grounds. First, it relies too heavily on chosen characterizations. Consider, for example, a situation in which the lawyer is seeking to obtain custody of the children for the husband, his client, in a divorce proceeding. The lawyer knows that the wife has been conducting an adulterous affair for some time and that, under the law of the jurisdiction, this would disqualify her from obtaining custody. He also knows that the wife would be humiliated by having details of the affair brought out in open court. If he chooses to advance his client's interests and raise the issue of adultery, is he merely raising a point of law (an institutional action) or is he doing a personal wrong to the wife? The first characterization would pass Fried's test; the second might not. No important moral principle should rest on arbitrary distinctions between equally appropriate characterizations of the actions.

Second, there is the more fundamental problem of absolving oneself of moral responsibility by shifting guilt to institutional shoulders. How can an action done within an institutional setting be morally appropriate, perhaps even morally required, when it is *wrong*? Fried suggests that responsibility shifts because the action is the action of the system—the institution—and not of the individual lawyer. Fried's reasoning seems to rest on the assumption that "if an action is essentially institutional then the good or bad consequences of the action are not attributable to the agent, but only to the institution." But, while it is important, for purposes of correction, to trace "institutional wrongs" back to the source, it does not follow that only the system may be blamed, or that the only possible action is to seek institutional reform. It is a mistake to insist that either the system is to blame or the individual agent is to blame. It is possible that moral criticism of both is appropriate. To rest blame with just one or the other suggests that when the system is to blame, there is nothing the individual can do about it but work to change the system, or the offending part of it. But that is not true: one can also avoid exploiting the defects of the law to the injury of others.

The lawyer must recognize that the institution acts only through the voluntary activities of the lawyer and client. The lawyer is not the instrument of the institution; rather the institution is the instrument of the client and the client engages the lawyer to make use of the instrument.

It is far more desirable to recognize at the outset that the lawyer, as well as the client, bears at least some responsibility for harms done by both "institutional" and "personal" actions. The question, then, is whether in particular cases there is a moral justification for the harms done. Whether there is or is not will be determined by the substantive moral considerations relevant in the case. And it is these substantive moral considerations that the responsible lawyer must take into account in making his or her decision.

Notes and References

[1]The concerns discussed in this essay were first suggested to me in discussions with several participants at the Institute on Law and Ethics sponsored by the Council for Philosophic Studies during the summer of 1977. Larry Alexander, Bernard Williams, and Gary Bellow were especially helpful. An earlier version of this essay was written as a background paper for the Philosophical Perspectives on Public Policy Project of the Center for Philosophy and Public Policy, the University of Maryland, College Park.

[2]For example: ''Lawyers, as guardians of the law, play a vital role in the preservation of society. The fulfillment of this role requires an understanding by lawyers of their relationship with and function in our legal system. A consequent obligation of lawyers is to maintain the highest standards of ethical conduct.'' ABA, *Model Code of Professional Responsibility,* Preamble, at 1 (1980) [hereinafter Code] (footnote omitted).

[3]For examples, see text accompanying notes 11–15 infra.

[4]T. Macaulay, ''Lord Bacon,'' in *Critical and Historical Essays* 2, (F. Montague, ed. 1903) 152.

[5]Quoted in Curtis, ''The Ethics of Advocacy,'' Stan. L. Rev. 4 (1951) 20.

[6]Id.

[7]See M. Freedman, *Lawyers' Ethics in an Adversary System* (Indianapolis, Bobbs Merrill, 1975) 27–42.

[8]See Code, supra note 2, Canon 4, especially EC 4-1, EC 4-5, DR 4-101(A), DR 4-101(B), DR 4-101(C)(3).

[9]See id. Canon 7, especially EC 7-27, DR 7-102(A)(4), (5), DR 7-102(B)(1). Also see ABA Project on Standards Relating to the Prosecution Function and the Defense Function § 7.7 (Approved Draft 1971).

[10]I borrow this term from R. Wassertrom, ''Lawyers as Professionals: Some Moral Issues,'' *Human Rights* **5,** (1975) 2–8.

[11]People v. Belge, 83 Misc. 2d 186, 372 N.Y.S.2d 798 (Onondaga County Ct.), aff'd mem., 50 A.D.2d 1088, 376 N.Y.S.2d 771 (1975), aff'd per curiam, 41 N.Y.2d 60, 359 N.E.2d 377, 390 N.Y.S.2d 867 (1976).

[12]See Code, supra note 2, EC 4-1, EC 4-4. Also see Callan & David, ''Professional Responsibility and the Duty of Confidentiality: Disclosure of Client Misconduct in an Adversary System,'' *Rutgers L. Rev.* **29** (1976) 332. The prosecution argued that failure to report the deaths amounted to a criminal violation of the New York State Public Health Law, 83 Misc. 2d at 187, 372 N.Y.S.2d at 799, which would render disclosure *permissible* under DR 4-101(C)(2). On more general grounds, however, it is hard to imagine that the ends served by the Health Code could outweigh the demands of confidentiality, if consideration of the much more significant injury to the families of the murdered women could not.

[13]242 F.2d 452 (7th Cir. 1957), cited in C. Fried, ''The Lawyer as Friend: The Moral Foundations of the Lawyer-Client Relation,'' *Yale L. J.* 85 (1976) 1064 n. 13.

[14]The *Zabella* court reasoned that "[o]f course, the jury was justified in thinking that defendant who then was in a position of some affluence and was the Chief Executive Officer of the Chicago Savings and Loan Association should feel obligated to pay an honest debt to his old friend, employee and countryman. Nevertheless, we are obliged to follow the law of Illinois." 242 F.2d at 455.

[15]See Code, supra note 2, EC 7-1, DR 7-101(A)(1).

[16]Perhaps one of the most serious general objections to Utilitarianism is that, although it professes to give full respect to all sources of value, it creates its simple normative structure by reducing all such values to a single dimension. The net effect is that either it distorts radically the world of human concerns, or it limits its scope to that range of values in which its simplifying assumptions are most natural.

[17]T. Nagel, "The Fragmentation of Value," in *Mortal Questions* (Cambridge: Cambridge Univ. Press, 1979) 134.

[18]See Nagel, supra note 17, at 135.

[19]See generally Aristotle, *Nicomachean Ethics,* bk. VI (H. Rackham trans. 1962).

[20]See S. Hampshire, *Two Theories of Morality* (Oxford: Oxford Univ. Press, 1977) 29–39; Nagel, supra note 17, at 135. See also Hampshire, "Public and Private Morality," in Hampshire, ed., *Public and Private Morality* (Cambridge: Cambridge Univ. Press, 1978), pp. 29–33.

[21]I borrow this example, for an entirely different purpose, from B. Williams, "Moral Luck," *Proc. Arist. Soc. Supp.* **50** (1976) 124.

[22]One aspect of the failure of professionals in law to appreciate the moral costs of their actions is captured by G. K. Chesterton in "The Twelve Men," in *Tremendous Trifles* (1955) 57–58:

> [T]he horrible thing about all legal officials, even the best, about all judges, magistrates, barristers, detectives, and policemen, is not that they are wicked (some of them are good), not that they are stupid (several of them are quite intelligent), it is simply that they have got used to it.
>
> Strictly they do not see the prisoner in the dock; all they see is the usual man in the usual place. They do not see the awful court of judgment; they only see their own workshop.

[23]See G. Haydon, "On Being Responsible," *Phil. Quart.* **28** (1978), 46–57.

[24]This appears to be the approach suggested by Wasserstrom. See Wasserstrom, supra note 10, at 12.

[25]A useful formal discussion of exclusionary reasons can be found in J. Raz, *Practical Reasons and Norms* (London: Hutchinson Univ. Press, 1975), pp. 35–48.

[26]By "conceptions of the lawyer's role," I do not mean some abstract model of a lawyer's professional behavior. Rather I have in mind the more or less complex pattern of beliefs and attitudes that tend to structure one's practical judgment and one's view of one's actions and relations to others, *i.e.*, the view of one's self in the role. Although there is a personal or idiosyncratic element in any person's conception, nevertheless, because the role of lawyer is largely socially defined, significant public or shared elements are also involved. I shall concentrate on these latter elements, keeping in mind that they are shared elements in an individual's self-conception in the role.

[27]Although my argument has general implications for the evaluation of conceptions of many professional roles, I shall restrict my attention here to what I shall call the standard conception of the lawyer's role.

[28]Samuel Johnson is often quoted with approval in support of this idea: " '[A] lawyer has no business with the justice or injustice of the cause which he undertakes, un-

less his client asks his opinion, and then he is bound to give it honestly. The justice or injustice of the cause is to be decided by the Judge.' " *Boswell's Journal of a Tour to the Hebrides,* Aug. 15, 1773, at 14 (F. Pottee & C. Bennett eds. 1936) (quoted in M. Freedman, supra note 7, at 51).

[29]J.-P. Sartre, "Existentialism" in *Existentialism and Human Emotions* (New York: Washington Square Press, 1957), pp. 9–51.

[30]See generally id. and J.-P. Sartre, *Being and Nothingness* (New York: The Philosophical Library, 1966) 86–116.

[31]E. Goffman, *Encounters* (Indianapolis 1961) 87–88.

[32]Consider, for example, the epitaph on a Scottish gravestone: " 'Here lies Tammas Jones, who was born a man and died a grocer.' " D. Emmet, *Rules, Roles and Relations* (London: Macmillan, 1966) 154 (quoting W. Sperry, *The Ethical Basis of Medical Practice* (London, 1951)) 41.

[33]I am indebted to Bernard Williams' lectures at the Institute on Law and Ethics sponsored by the Council for Philosophical Studies in 1977 for the remarks at this point.

[34]Curtis, supra note 5, at 22 (1951) [quoting R. Ellman, *Yeats* (1949) 178].

[35]See G. Bellow and J. Kettleson, "The Mirror of Public Interest Ethics: Problems and Paradoxes," in *Professional Responsibility: A Guide for Attorneys* (ABA, 1978), pp. 257–258:

> At the root of the dilemma is a professional ethic that requires a sharp separation between personal and professional morality. The lawyer is asked to do "as a professional" what he or she would not do "as a person"; to subordinate personal qualms about results in particular cases to the general rule of law and the bar's role within it. There is much to be said for such a combination of responsibility and neutrality, if the "law job" is to be performed. But it may be that over time, such a division between the personal and the professional will atrophy those qualities of moral sensitivity and awareness upon which all ethical behavior depends.

[36]See Elkins, "The Legal Persona: An Essay on the Professional Mask," *Va. L. Rev.* 64 (1978) 749, which argues that the pressure upon lawyers to identify with their role comes from a number of pervasive factors, including linguistic factors, specialized modes of reasoning, and even characteristic clothing styles.

[37]As a result, serious questions arise when the lawyer acts negligently or irresponsibly. Is the client, thereby, committed to the consequences of such actions, of which the client may not have been aware or did not approve? In general, the client is committed; failure of counsel to appear or respond may result in a default judgment as effectively as if the client never retained counsel in the first place. See L. Mazor, "Power and Responsibility in the Attorney-Client Relation," *Stan. L. Rev.* **20** (1968), 1121–23, 1124 and n.24.

[38]Wassertrom, supra note 10, at 14.

[39]See generally W. Simon, "The Ideology of Advocacy: Procedural Justice and Professional Ethics," *Wis. L. Rev.* 1978 3, 96; Code, supra note 2, DR 7-106(C)(4) (attorney prohibited from expressing personal opinions regarding, *inter alia,* the justness of a cause).

[40]See note 35 supra.

[41]Cf. A. Neier, *Defending My Enemy* (1979) (belief in the higher value of the legal system *qua* system necessitates defending persons and causes antithetical to the lawyer's own beliefs).

[42]This point was suggested to me by Philippe Nonet.

[43]See generally R. Dworkin,*Taking Rights Seriously,* (Cambridge, Mass.: Harvard Univ. Press, 1978), H.L.A. Hart, *The Concept of Law* (Oxford: Oxford Univ. Press, 1961) pp. 199, 205–207; D. Richards, *The Moral Criticism of Law* (Encino, Calif.: Dickenson, 1977), pp. 31–36.

[44]See text accompanying note 34 supra.

[45]See text accompanying notes 16–20 supra.

[46]This may explain, in part, the attitude of "ethical minimalism" among lawyers which many, both within and outside the profession, deplore. This minimalism is an understandable reaction, in light of the fact that there are few fixed and settled rules in the Code and the lawyer is effectively cut off from the resources needed to resolve the indeterminacies unavoidably left by the Code.

[47]See pages 41–42 supra.

[48]B. Williams, "Politics and Moral Character," in Hampshire, *Public and Private Morality,* supra note 20 at 64. Milgram's well-known experiments underscore the commonplace that the more we are able to distance ourselves (often literally) from the consequences of our actions, the more we are able to inflict pain and suffering on others without moral qualms. See generally S. Milgram, *Obedience to Authority: An Experimental View* (New York: Harper and Row, 1974), pp. 32–43.

[49]Fried, supra note 13, at 1071–72.

[50]Tammas Jones, see note 32 supra, was not just a grocer; he was also, *inter alia,* a father, husband, friend, and neighbor. It was possible for him to relate to his family, customers, neighbors, and friends, not as a role-agent, but as a person, because it could have been recognized that his moral personality penetrated through his activities in his roles, and that these roles did not exhaust that personality.

[51]E. Goffman, supra note 31, at 142.

[52]David Hoffman, a nineteenth century legal educator in Maryland, offered a conception of lawyering in which the lawyer's sense of responsibility was central. D. Hoffman, *A Course of Legal Study* 2 (2d ed. Baltimore, 1836) (I am indebted to Michael Kelly for this reference.). Hoffman wrote: "My client's conscience, and my own, are distinct entitites: and though my vocation may sometimes justify my maintaining as facts, or principles, in doubtful cases, what may be neither one nor the other, I shall ever claim the privilege of solely judging to what extent to go." Id. at 755. Furthermore, he insisted that:

> Should my client be disposed to insist on captious requisitions, or frivolous and vexatious defences, they shall be neither enforced nor countenanced by me . . . If, after duly examining a case, I am persuaded that my client's claim or defence . . . cannot, or rather ought not, to be sustained, I will promptly advise him to abandon it. To press it further in such a case, with the hope of gleaning some advantage by an extorted compromise, would be lending myself to a dishonourable use of legal means, in order to gain a *portion* of that, the *whole* of which I have reason to believe would be denied to him both by law and justice.

Id. at 754.

[53]S. Kadish and M. Kadish, *Discretion to Disobey* (Stanford: Stanford Univ. Press, 1973), pp. 31–36.

[54]See id. at 33–34.

[55]Id.

[56]Id. at 35.

[57]Id. at 35–36.

[58]See Fried, supra note 13, at 1066 n.17.

[59]See M. Freedman, "Personal Responsibility in a Professional System," *Cath. U. L. Rev.* 27 (1978), pp. 193–95.

[60]Code, supra note 2, DR 2-109(A)(1).

[61]Id. DR 2-109(A)(2).

[62]Id. EC 2-30.

[63]Id. EC 2-26.

[64]See id. DR 2-110(C).

[65]Id. DR 2-110(C)(1)(a). Withdrawal is permitted for a number of other reasons, all unrelated to the present argument. These include, among others, failure by the client to pay fees, DR 2-110(C)(1)(f), and inability to work effectively with co-counsel, DR 2-110(C)(3).

[66]Id. DR 2-110(C)(1)(b), (c).

[67]Id. DR 2-110(B)(1). Withdrawal is mandated for a variety of morally neutral reasons, such as a conflict of interest with another client whom the lawyer is representing, DR 2-110(B)(2); see DR 5-105, or the lawyer's ill health, DR 2-110(B)(3).

[68]Id. EC 2-30.

[69]The Code *does* allow permisive withdrawal, however, if the client insists that the lawyer engage in conduct that is contrary to the lawyer's advice and judgment *and the matter is not before a tribunal*. Id. DR 2-110(C)(1)(e).

[70]Fried, supra note 13, at 1073.

[71]Id. at 1075.

[72]Id. at 1073.

[73]Id. at 1082–86.

[74]Id. at 1084–86.

[75]Id.

[76]Id. at 1085.

[77]Id. at 1086.

Professionals, Clients, and Others

Michael D. Bayles

Professionals frequently claim to be exempt from certain obligations that nonprofessionals have and to possess others that nonprofessionals do not. The most blatant forms of dissociation of professional obligations from those of nonprofessionals, especially in the medical field, have generally been abandoned under the onslaught of recent criticism.[1] The remaining areas in which professionals claim to be exempt from ordinary morality are complex and debatable. Most discussion of this general issue has focused on obligations of professionals to their clients. Perhaps even more important are professionals' obligations to persons other than their clients—third parties. This paper is limited to consulting professionals who, because of their special relations with clients, present the strongest claims for exemption from ordinary morality.

The traditional professional position is that so long as conduct is legally permissible, obligations to clients take precedence over ordinary obligations to others.[2] Because of their obligations to clients, proper conduct for professionals often differs from that for nonprofessionals. The frequency with which the following comment of Lord Brougham is approvingly quoted testifies to the strength with which this view is held.

> . . . An advocate, in the discharge of his duty, knows but one person in all the world, and that person is his client. To save that client by all means and expedients, and at all hazards and costs to other persons, and, amongst them, to himself, is his first and only duty; and in performing this duty he must not regard the alarm, the torments, the destruction which he may bring upon others. Separating the duty of a patriot from that of an advocate, he must go on reckless of the consequences, though it should be his unhappy fate to involve his country in confusion.[3]

Taken literally, Brougham's comment would be the strongest possible claim to violate ordinary morality, for he does not even restrict the means used to legal ones. In the original context of a thinly veiled threat to expose the King's adulterous affairs should charges be brought against his client the Queen, Brougham implicitly restricted the means to legal ones. Almost everyone agrees that professionals should restrict themselves to legal means, except for rare acts of civil disobedience such as courtroom disruptions.[4]

The problem of balancing obligations to clients and to others may be broken into more discrete issues. One issue is how the *law* should balance professionals' obligations to clients and to others. Should accountants be legally required to report cases of fraudulent corporate balance sheets? Should engineers be legally required to report poorly designed products to consumer protection agencies? Ought lawyers to be legally required to report confidential information received from clients, such as where murderers have left their victims' bodies or their clients' intentions to commit new crimes? Simply stating that professionals must act within the law provides no guidance about what the law should be.

A second issue concerns balancing obligations when the conduct is within legal limits. Should professionals assist clients in achieving legally permissible ends that are contrary to ordinary morality? Can any legally permissible means be used to assist clients, regardless of their conformity to the dictates of ordinary morality? Since the conduct of lawyers in assisting their clients almost always affects others, whereas the conduct of other professionals is less likely to do so, these issues are more pressing for lawyers. Nonetheless, all consulting professionals occasionally face them.

One difficulty in analyzing these problems concerns the basis for claiming that ends or means are contrary to ordinary morality. If they are illegal, at least a community has clearly judged that they ought not be pursued or done. If they are not illegal, in a pluralistic democracy no generally accepted judgment of their morality may exist. However, the question is not whether the conduct is contrary to what most people believe to be wrong, but whether it is contrary to a rationally defensible morality applying to nonprofessionals. Furthermore, this aspect is often not in question because the conduct is detrimental to others and thus violates an ordinary moral norm not to harm others. In a few cases whether harm occurs may be disputed, but it does not affect the issue under consideration—whether professionals are sometimes exempt from ordinary morality, whatever it may require.

A couple of brief examples indicate the type of problems involved in balancing ordinary moral obligations to others with special ones to clients. Suppose a consulting engineer discovers a defect in a structure that engineer knows the owner is about to sell. If the owner will not disclose the defect to the potential purchaser, ought the engineer to do so? Suppose a lawyer learns that his or her client intends to commit perjury on the witness stand. Should the attorney report the client's intentions to the authorities? May the lawyer morally put that client on the stand? Suppose a lawyer's client wishes to plead the statute of frauds to a debt the lawyer knows was in fact incurred by the client. Ought a lawyer to interpose technical defenses to what a client admits is a morally just claim? May a lawyer destroy the credibility of a vulnerable witness whom that lawyer knows is telling the truth?

Charles Fried contends that a lawyer may morally assist a client to achieve any legally permissible end. He views the lawyer as a limited friend of the client. Like other friends, a lawyer may give special consideration to the interests of the client over those of others and do things for the client that the lawyer would not do for him or herself. Fried assumes that the legal system is reasonably just. Given that the client's conduct is legal, he claims that for the client to be unable to achieve desired ends because of ignorance or misinformation about the law would violate the client's rights. Consequently, assisting clients to achieve their legal rights is, he claims, "always morally worthy."[5] "The lawyer," Fried contends, "must distinguish between wrongs that a reasonably just legal system permits to be worked out by its rules and wrongs which the lawyer personally commits."[6] The lawyer ought not commit personal wrongs. In lying to a judge or abusing a witness, the lawyer acts in a personal way, whereas in asserting a statute of limitations to defeat a legal claim, a lawyer is acting as a representative because that act is a legally defined one. When a lawyer acts as a representative and not personally, he or she is insulated from responsibility.[7]

Several difficulties confront Fried's view. A major one is distinguishing between representative and personal acts. Fried suggests that the distinction is based on how the acts are defined. Filing motions and making objections are acts defined by the legal system, whereas lying and abusing persons are not so defined, but have meaning independent of the legal system. This distinction is unclear; lying to a judge may be definable independent of the legal system, but the act may also be described as committing a fraud upon a tribunal. A judge is defined by a position within a legal system, so that even 'lying to a judge' is not completely specifiable independent of a legal system. Humiliating a witness may also be described as cross-examining a witness, which is a legally defined act. Conduct may be properly described in more than one way, and some descriptions of an act may refer to a legal system and others not. One suspects that Fried is relying heavily on his assumption that the act does not violate the law or norms of professional ethics.[8] The American Bar Association *Code of Professional Responsibility* prohibits an attorney asking a question "that is intended to degrade a witness or other person."[9] Thus, Fried's example of humiliating a witness is already prohibited by this assumption.

Even if it can be made, the distinction between wrongs a system permits and those done personally has limited applicability. It best applies to an attorney in a trial situation. In counseling or negotiating, fewer of an attorney's acts are defined by reference to the legal system. Indeed, since negotiation also occurs completely outside a legal context, probably no acts of negotiation are necessarily defined by the legal system. The distinction is also perforce restricted to lawyers. The acts of accountants and most other professionals are not defined by a system of rules as are some of those of lawyers. The analysis will not apply to professional ethics generally, or if it is applied, will confine other professionals to all the requirements or ordinary morality.

More fundamentally, even granting the distinction, why should the way an act is defined have the conclusive weight Fried gives it in determining the morality of conduct? Presumably, the reason is that the legal system allocates rights in order peaceably to settle disputes and that objections to injustice the system permits should be corrected through the political process. But many moral constraints limit conduct that is not beyond the realm of legal tolerance; it would be a morally deficient society in which everyone pursued personal claims to the extent of the law.

Fried may contend that while this is true, the fault lies with the client, not the attorney. Although this reply indicates that the client is immoral, it does not necessarily absolve the lawyer from responsibility. Fried agrees, but tries to limit the lawyer's responsibility to those personal acts done on the lawyer's own initiative rather than as strictly representative of the client. This leads back to the problems with the original distinction. Disregarding them, no argument is given why acts performed in a representative capacity are subject only to legal constraints and not also moral ones. Even friends recognize moral constraints upon what they will do on behalf of each other. In the end, Fried's analysis places legal rights above moral considerations, for he contends that not providing a client full legal rights is wrong, even if providing them requires or involves immoral but legal means. No reason exists to so exalt legality.

Richard Wasserstrom points out the difficulties that ensue from distinguishing, as Fried does, between the conduct of a person in a professional role and as a private citizen. He notes "costs" involved in ascribing an amoral character to a lawyer's activities in his professional role.[10] Arguments such as Fried's in favor of allowing such conduct assume that the legal system is just. To the degree the legal institution is not just or wise, such role-differentiated norms may be undesirable. A lawyer's character will also be adversely affected. As a lawyer, he or she needs to be competitive, aggressive, ruthless, and pragmatic. Such traits cannot be readily confined to the professional role. These character traits usually affect most conduct and become relatively permanent features of the personality.

At this point a principle is needed to resolve conflicts between a professional's obligations to clients and to others. Just as religious obligations are limited by ordinary morality and law when they adversely affect persons outside the sect, so professional obligations to clients must be limited by moral obligations to others. A balancing of these obligations cannot be accomplished solely on a case-by-case basis, for detrimental effects of general professional practice do not result from single cases. In short, the issue is how one determines the norms of professional roles. This issue cannot be settled within the perspective of professionals. Reference must be made to the broader framework of social values.

The approach here is to adopt a consumer perspective and consider what balance would most promote the values of a democratic society. In particular, the proposed test is to ask what balance of obligations a reasonable person who may be in the role of client or affected person would find best promotes a society with the values of freedom, equality, prevention of harm, welfare, and privacy. The balancing is between what a person would want done for himself as a client and the effect of such conduct upon him as an affected party. The proposed test subjects professionals to obligations of ordinary morality. It evaluates professional obligations by the values of society and ordinary citizens, yet ordinary citizens may make allowances for the particular function professionals play in society.

There are three important steps to the procedure. First, one must identify and comparatively weigh the values of the client against those of others who will be affected. Unlike a standard utilitarian analysis, this evaluation is not done simply on the basis of the happiness of the persons affected. Second, one must consider the general probability of one being in either position—of client or affected party. This consideration is especially important for issues in which many persons may be affected. Third, one must remember that one is considering rules for professional roles; one is not deciding a particular case and is not able to give a different answer for another similar case at a later time. In general, the procedure is designed to foster asking whether one would rather live in a society with professionals governed by one set of norms or by another. Consideration of examples may clarify the procedure.

The simplest case is that in which only three parties are involved—the professional, the client, and one affected party. The previous example of a consulting engineer who has found a structural defect in a building which the client, the building's seller, has not revealed to the prospective buyer is of this type. The client's values involved are confidentiality and financial interests. The engineer's report was confidential, and should the prospective purchaser learn of the defect, no doubt that buyer will not be willing to pay as much. The prospective purchaser's affected values are financial, but may also be physical safety. If the defect makes the premises unsafe, then the buyer or the buyer's employees may be injured or killed should they occupy them without correcting the defect. To apply the recommended procedure, one should ask whether one would be willing to risk the financial loss and injury for financial gain and confidentiality. Because of the importance of physical safety, a reasonable person would conclude that knowledge of the defect as a purchaser is more important. Therefore, the reasonable person would support a norm whereby the engineer is obligated to inform the prospective purchaser of the defect.

To show the significance of the weighing of values, one may compare this case with one in which the engineer has determined that modification of the structure for the owner's desired use would be prohibitively expensive. In this case, a prospective buyer would not risk physical safety; however, should the buyer learn of the engineer's report, that buyer would have a bargaining advantage over the client. The values to be compared are financial interests and confidentiality for the client and financial interests for the prospective purchaser. A reasonable person would prefer a norm by which the engineer respects confidentiality in this case. The purchaser is quite capable of arranging protection from financial loss since no more need be offered for the structure than it is worth, even though the seller might have been willing to sell for less.

More complex cases arise when a large number of persons may be affected by a client's conduct. In these cases, the values of the affected persons must be weighted by the number of persons involved. One way to do this is to weight the values of the general probability of being a client or affected party. An example may clarify this point. Suppose an accountant learns that the balance sheet of a corporation, while done according to accepted accounting procedures, gives a misleading impression of the financial strength of the corporation, making it appear to be in a better condition than it actually is. The management's values are financial interests, including job security, and confidentiality. The affected parties include all stockholders as well as potential stock purchasers. Although the financial loss to management of the actual financial position becoming known is probably greater than for any particular stockholder, there are many more stockholders. Consequently, to apply the recommended procedure, one must use the general probability of being a stockholder versus that of being a member of management. Since the ordinary person is much more likely to be a stockholder, at least through a retirement system, that ordinary person is apt to prefer a norm requiring disclosure to one requiring the preservation of confidentiality. Thus, an accountant should break confidentiality in this case.

Perhaps the most discussed issue involving the balancing of a professional's obligation to keep client communications confidential and ordinary obligations stemming from public values is whether criminal defense attorneys ought to inform on clients who commit perjury. Monroe H. Freedman contends that the lawyer's responsibility of confidentiality overrides the public's values of conviction of criminals and avoidance of perjury. He maintains that if a lawyer knows the client is going to commit perjury, the lawyer ought not reveal either the client's intention to do so or that the client has already done so.[11] Freedman emphasizes the problems of alternatives. If the attorney does not guarantee confidentiality, then the client will not confide in that attorney. If the attorney seeks to withdraw from the case or to ignore the testimony in making the case's arguments, the judge or jury will infer that the client lied.

Freedman's position is apparently contrary to the ABA *Code,* which at least permits, and by interpretation requires, that a lawyer inform authorities if he or she knows beyond a reasonable doubt that the client intends to commit a crime, that an attorney not use perjured testimony, and that a person or tribunal against whom a fraud has been perpetrated be informed.[12] Nonprofessionals arguably have an obligation to report such conduct and even if they do not, it is surely permissible for them to do so. Thus, Freedman's view clearly places professionals under a different set of norms than the ordinary ones.

Freedman assumes a full-blown adversary model of the legal system. Although that model fits the criminal trial context better than others, even there it need not be unmitigated by considerations of ordinary morality. Such considerations may limit confidentiality by prohibiting the performance of, or assistance in, acts that are unethical or illegal. The basic issue is the extent to which the adversary model should be applied even in criminal cases, and Freedman's assumption begs that question.

By the proposed test, the question is whether a reasonable person would forgo the increase in convictions that would result if lawyers were obligated to inform in order that this same reasonable person be able to commit perjury without being informed on by that lawyer. Many people might immediately respond that lawyers should inform on clients, because they would never be criminal defendants and have nothing to fear from such a policy. That is not the appropriate test. One must instead imagine that one is choosing between the values for oneself, that one must sacrifice one value for the others. Thus, the procedure was specified as using the general probability of one being in positions, that is, the general probability of a member of the population being a criminal defendant.

Given that clarification, one must realistically examine the consequences of alternative policies. Requiring lawyers to inform on clients will only slightly increase convictions. Most convictions are guilty pleas to reduced charges. That also means that few occasions for perjury exist. Few people dispute that an attorney in a civil case should reveal a client's perjury. The reason may be either that a specific person is injured by the testimony or that the defendant has less at stake—usually only financial interests rather than liberty. Although the effects may be less direct and affect unidentified individuals, the harm from criminal offenses is surely greater than from civil cases, which counterbalances the greater loss to the defendant. Nor does it follow that a defendant can get a mistrial by simply committing perjury. An attorney might be required to complete the trial (either arguing or not arguing a defendant's perjured testimony) and to inform authorities afterwards for purposes of prosecution.

The value choice is avoidance of conviction of the innocent and adequate legal defense versus conviction of the guilty and preservation of the legal system from corruption. Since lying may not significantly help prevent conviction of the innocent, a reasonable person may hold that lawyers should inform. Such a system might mean that more clients would lie to their attorneys, but the alternative is to permit attorneys to knowingly assist clients in lying to the court. No clear reason exists for a reasonable person to prefer the latter to the former.

The discussion herein has not attempted to distinguish between those professional norms that should be legally enforced and those that should be left to the realm of other techniques. To make that distinction, one needs to consider such matters as costs, effectiveness, and fairness of enforcement. Instead, the purpose herein has been to illustrate an intellectual procedure for determining the substantive content of professional norms. The point of considering these specific cases has not been to provide definitive solutions to them, but to illustrate the proposed test for balancing professionals obligations to clients with those to others. Professionals are not completely exempt from the constraints of ordinary morality, but the functions of their professions in society must be taken into account in the balancing process. Even then, the balancing is not to be done on the basis of the values peculiar to particular professions. Instead, the balancing is done from the perspective of a reasonable non-professional in society judging by justifiable democratic values. Consequently, even the special weight professionals may give to client interests is ultimately justified by considerations of ordinary morality. The dictates of ordinary morality apply to everyone, but the various situations and roles of people in society may require somewhat different balancing for the sake of achieving democratic values for ordinary citizens.

NOTES AND REFERENCES

[1]Such matters concern the permissibility of lying to dying cancer patients about their condition and experimenting without informed consent.

[2]According to at least one author, the American Bar Association's *Code of Professional Responsibility* holds that lawyers owe clients almost unqualified loyalty and others only what is compatible with that duty. Geoffrey C. Hazard, Jr., *Ethics in The Practice of Law* (New Haven: Yale University Press, 1978), p. 8.

[3]Quoted in Monroe H. Freedman, *Lawyer's Ethics in an Adversary System* (Indianapolis: Bobbs-Merrill, 1975), p. 9.

[4]See Jethro K. Lieberman, *Crisis at the Bar: Lawyer's Unethical Ethics and What to Do About It* (New York: Norton, 1978), pp. 168–169.

[5]"The Lawyer as Friend: The Moral Foundations of the Lawyer-Client Relation," in *1977 National Conference on Teaching Professional Responsibility: Pre-Conference Materials,* Stuart C. Goldberg, ed., (Detroit: University of Detroit School of Law, 1977), p. 144; Charles Fried, *Right and Wrong* (Cambridge: Harvard University Press, 1978), pp. 181–182.

[6]"Lawyer as Friend," p. 153; *Right and Wrong,* pp. 191–193.

[7]*Right and Wrong,* p. 183.

[8]"Lawyer as Friend," p. 151.

[9]DR 7-106(C)(2).

[10]"Lawyers as Professionals: Some Moral Issues," in *1977 National Conference, op. cit.,* pp. 116–118.

[11]*Lawyer's Ethics,* Chap. 3, esp. pp. 27, 40.

[12]DR 4-101(C)(3), ABA *Opinion* 314 (1965); DR 7-102(A) (4); and DR 7-102(B) (1).

Part II

Business Ethics

Business Ethics

An Introduction

Joseph Ellin

The essays in this section have a common theme. They all reject the contention that there is in some way a special—and from the ordinary point of view far less demanding—morality that applies to business and to people in the conduct of business. The essays examine the contention that business managers are, and properly should be, motivated by a single overriding consideration: the pursuit of monetary gain. On this theory, market and legal considerations aside, businesspeople are allowed to ignore the effects their actions might have on everyone (employees, consumers, the general public), and make their decisions solely on the basis of calculations designed to maximize profit. Typically, of course, the market will require that one's actions be fair: an employer who does not pay adequate wages will not get competent employees, a manufacturer who does not charge reasonable prices will lose business to competitors, and so on—but this is really irrelevant from the point of view we are now considering, since it assumes that whether or not one's actions are fair, decent, moral, or just should have no bearing on one's business decisions. Only what is profitable counts.

There is, however, one constraint that business must consider, and that is the law. Society has decided that certain things (for example, paying wages below minimum) must not, and other things (affirmative action programs) must be done; obedience to the laws embodying these decisions is required by the business morality we are considering. There is no right of conscientious objection to legal mandates. Beyond the law—and, of course, the market—however, business has no moral obligations. Whatever is not prohibited by law is allowed; whatever is not required, need not be done. In the first essay in this section, L. Duane Willard examines this contention that action within the law is morally sufficient in business. There are, he points out, many reasons that lead certain authors to adopt this view, ranging from the alleged uncertainty and disputability of extralegal rules of morality, to the competitive disadvantage a person in business would place on him or herself should he or she adopt (in the name of morality) self-restraints more stringent than those adopted by competitors. "Do unto others before they do unto you," though a gross and unacceptable principle from the standpoint of ordinary morality, is the only prudent maxim in the dog-eat-dog world of business competition.

Willard is not persuaded. "Operating in a free enterprise system does *not* mean by definition operating free of all moral constraints beyond the law." Although there is no moral problem with wanting to make a profit, and indeed with wanting to make as much profit as one can, there is a moral problem with wanting to make a profit at the expense of harming other people. Furthermore, Willard points out that the excessive pursuit of profit ("greed," to give it its proper name) is harmful to the greedy person as well as to others affected by his or her actions. Greed causes a person to develop a narrow, inhibited view of life, and to become a slave to success and money. Greed has a destructive effect on health, friendships, and family. "If the businessperson insists upon satisfying the desire for more and more profit . . . no doubt it can often be done . . . But let us be done with the rationalizing excuse that such a businessperson is not doing anything immoral, that the principles and rules of ordinary morality do not apply in his or her case."

Robert V. Hannaford continues the analysis of "the bottom line mentality" by tracing it to the argument usually attributed to Adam Smith: If each pursues his or her own self-interest, the consequence will be the good of the whole. Smith's defense of this proposition with respect to economic activity is the philosophical foundation upon which such contemporaries as Milton Friedman argue that the only responsibility of business is to maximize profits. A business executive who fails in this responsibility, and allows other social aims to divert him or her from the path of self-interest, according to Friedman in a number of books and articles, undermines the free market, endangers liberty, and paves the way for the restrictions of socialism. All these accusations Hannaford vigorously disputes. Hannaford suggests that what really threatens freedom is the "single-minded pursuit of profit." "Acting to fulfill social obligations is not contrary to a firm's business interests. Business interests include community interests. Corporate decisions should center on what is in the total interest of the corporation. If a corporate decision is well-made, it will be seen to include concern for the community that makes the corporation possible.''

Hannaford diagnoses the basic mistake of theorists like Friedman as confusing self-interest with selfishness. (This confusion is also pointed out by Koehn in the essay following Hannaford's.) But people are interested in other things besides their personal financial gain. Farmers, for example, may sacrifice some profit to avoid polluting their neighbor's water supply. In thus acting freely, they are showing that they have an interest in social responsibility, and presumably are gaining some satisfaction therefrom. Though their actions are not *dictated* by the market and its price mechanism, it can hardly be said that by their responsible action they are undermining that market or the freedom of the greater society. (As Koehn points out in his essay, when Friedman uses the term ''self-interest,'' he means the pursuit of whatever a person values. Since some people may value responsibility over profit, an ethic of self-interest does not *necessarily* require us to be single-minded profit maximizers.)

A major point made by Friedman is that corporate managers are responsible to their stockholders, and ought to use stockholders' money only in ways that stockholders approve of, that is, presumably, for the most profitable investments. To this, Hannaford points out that it is not at all clear what stockholders do want. It remains to be proved that stockholders want the greatest profit, no matter what the cost in social responsibility. Do stockholderes in public utilities really want the greatest return, even if this means shoddy construction and unsafe conditions at nuclear generating plants? Is this really in their interest? Hannaford closes by suggesting that truly responsible business decisions are made in accord with ''what the common good requires.'' ''Business must not only acknowledge its social responsibility, it must often bear an extra burden of responsibility.'' To the extent that business denies its social responsibility, everyone else is encouraged to deny their own responsibility. This growing lack of responsibility, showing a failure to respect other people, Hannaford finds to be the real threat to our free society.

The essay by Donald R. Koehn puts the views of Friedman in the context of similar views expressed by Hirshleifer and Ayn Rand. These writers all concur that the unrestricted pursuit of self-interest is beneficial to all, and hence requires no moral restraints. This view Koehn calls the ''tough capitalistic ethic,'' to distinguish it from the ''soft capitalistic ethic,'' represented in his essay by a former official of Sears Roebuck, which holds that the interests of everybody ought to be taken into account when executives make business decisions. According to the "tough" ethics, the soft view fundamentally misunderstands the nature of capitalism, which is a system in which the only restraints upon self-interest, apart from the law, are the restraints imposed by the market. According to this *laissez-faire* ethic, the only role of the government is to maintain the free market, and provide for the proper allocation of cost externalities and collective cost goods. All of this is justified by the wonderful self-direction of the market, which distributes resources, goods, and services without the direct intervention of anyone, thus allowing maximum freedom from external control and interference.

Although Friedman does not explicitly state an ethic of market behavior, analysis of his writings suggests that he holds what Koehn calls the "weaker version" of the tough market ethic, namely, that wholly selfish behavior, free of any restraints but those imposed by law and the market itself, is permitted, though not required. That people in business morally *must* act in a uniformly profit-maximizing way, unless restrained by law, is the "strong" version of the tough ethic. It is this version that Hershleifer seems to hold. On the weak view, there is nothing wrong, for example, with a business practicing racial discrimination, if it is profitable to do so, to the extent that the law allows. But on the strong version, presumably, a business *ought* to practice discrimination—is morally at fault if it fails to practice discrimination—when required by profit maximization. Now Friedman does not actually go this far (neither does Hershleifer explicitly), but it is a bit unclear what reasons he could have for refraining from so doing. Does not business have a positive obligation to maximize profit, and to do whatever it sees as necessary in pursuit of this single goal? But Friedman shies away from this explicit consequence.

The justification of the "tough" ethic is simply that by following it, and allowing the market to operate, the business person will in the long run produce more good (in terms of want satisfactions for society as a whole) than if he or she practiced self-restraint in accordance with some principles of morality. In section VI of his essay, Koehn confronts this assertion head on. By means of examples, he points out that this assumption of the tough ethic is simply fallacious. We can easily imagine cases in which the pursuit of self-interest imposes various harms on affected parties, without producing any corresponding greater good, either for the larger society or for the harmed individuals. Furthermore, in our actual economy, which is characterized by oligopolies or monopolies and in which avoidance of market constraints is by no means impossible, harmful practices (for example, price-gouging, built-in obsolescence) that an ideal free market would be expected to eliminate may prove quite profitable. Government action may be necessary to protect the public against unscrupulous promoters and others who are able to avoid free market constraints. Even the market itself requires *some* self-restraint; without it, we would have the kind of "free market" operation that characterizes not business, but organized crime. Therefore, Koehn concludes, markets are not "sufficient generally to convert selfishness to the service of others," and therefore it is not true that there is a special, less stringent ethic that applies to business.

The short essay by Norman C. Gillespie, while making clear that being in business does not excuse a person from following the precepts of ordinary morality, discusses conditions under which our morality would allow exceptions to its general rules. There seem to be two such circumstances: first, when nobody else is doing what they ought to do and therefore nothing would be accomplished if you did what you ought to do; and second, where, should you do what you ought to do, you would open yourself to be taken advantage of by others less scrupulous than yourself. An example (not from business) of the first case is jury duty: if no one else reports for jury duty, I would be wasting my time and accomplishing nothing were I to report; therefore I can properly excuse myself. An example of the second case would be industrial espionage: where everyone else is spying, it might be necessary to "fight fire with fire" in order to protect yourself against competitors' predations. These exceptions to a moral principle do not show that there is a special ethic for business; such exceptions are part of our ordinary morality, which is simply being applied to business. And there is another important point often overlooked by the "moral conventionalists" who advocate a special loose ethic in business. This is that in such situations, nobody is doing what they ought to do. It is precisely because everyone else is violating morality that *you* may be excused if you do so; the point being that you ought therefore to do whatever you can to improve the general moral situation. Conventionalists, anxious to show that *everyone* has a right to cheat, overlook that the right to cheat is conditional: *no one* has such a right, unless other people are already cheating (or may be expected to do so). Therefore, while cheating may be permitted, it is still obligatory to work for change. "At the very least," Gillespie argues, "executives should not thwart the impetus for change on the ground that business sets its own ethical standards."

The specific subject of bluffing (or lying) in business is discussed in the paper by Thomas L. Carson and Richard E. Wokutch. Bluffing occurs typically in the context of negotiations. After some discussion of what constitutes lying and what does not, the authors conclude that "bluffing generally does involve lying." Bluffing and other forms of deception are often defended on various grounds, for example, that the practice is profitable, that it is economically necessary, and that everybody does it. Carson and Wokutch concede that the market cannot be counted on to discourage deception by rendering it unprofitable in the long run; those who wish to beat the market by deceptive practices will be able to discover opportunities to do so. Nevertheless, the fact that a practice is profitable does not make it justifiable. "Economic necessity" would be a stronger justification, but "a firm that needs to practice lying or deception in order to continue in existence is of doubtful value to society." Such a firm should probably be allowed to go out of business: ". . . the long-term consequences of the bankruptcy of a firm that needs to lie in order to continue in existence would be better or no worse than those of its continuing to exist." There are two exceptions to this general truth, however. One occurs when in fact the consequences of the firm's going out of business *would* be very bad; in that case lying to save the firm might be allowed, since one is allowed to lie where necessary to prevent great harm. The other exception occurs when the persons being lied to "have no right to the information in question." Hence it would not be right to lie to customers in order to stay in business, but it might be right to lie or deceive one's bargaining adversaries in order to make the good deal that allows you to stay in business.

How about the argument that lying and deception are standard business practices? "The mere fact that something is standard practice or generally accepted is not enough to justify it." Nevertheless, in a situation in which someone is attempting to deceive *you,* you may be justified in using deception against that party: this is another example of the "fighting fire with fire" condoned by Gillespie in the previous essay. As Carson and Wokutch point out, in situations in which bluffing is the standard or expected practice (such as negotiations), there arises a strong presumption that the opponent *is* trying to deceive you, and hence deception on your part may be justified.

Since deception can be economically advantageous, the authors attempt to offer guidelines to the individual "who is concerned about the morality of lying and bluffing, but is also concerned about furthering his or her own economic welfare." These "guidelines"—which embrace advice to the public as well as to people in business—include explicit attention to the ethical issues involved in various negotiating tactics, strict legal definitions of what constitutes deceptive practices, and the development of financial and other incentives for honesty in economic transactions.

The final essay in this section takes a different approach to ethical problems in business. Not content to "justify" ethics, the author addresses the problem of what prevents ethical behavior in the business world. Society, according to Richard T. De George, is changing the mandate it has given to business. Financial accounting is no longer sufficient; society now demands "social accounting" as well. The problem is that the traditional model of the corporation makes it difficult for business to respond to the new social demands. Consequently, "what is needed is both a different image of the corporation and corresponding organizational changes." The traditional model of the corporation saw it as an organization constituted for certain limited ends—profit, production, provision of services—in which each corporate official plays a role determined by these ends. Moral responsibility as such does not apply to the corporation so conceived. Since corporations are formed to achieve given ends, they should be free, apart from constraints imposed by law, to act for those ends; society has no right to impose extraneous demands in the form of morality.

DeGeorge dismisses this model as "an ideology that is now in the process of erosion," one that is becoming growingly irrelevant in a world in which people are increasingly interdependent. Corporations are social entities that exist not through inherent right, but because they serve society's purposes. The ends they serve are society's ends, not ends of their own choosing. Although corporations as such may be free of moral responsibility, the people who run corporations are no less bound by morality than anyone else, and they can be held accountable by society for their actions.

The time has come, De George thinks, to speak of a new form of capitalism, "humanistic capitalism," a new view of business and a new theory of the corporation according to which those who run business

> . . . should be concerned for the people who work in business and for those affected by it. The time when people could be considered simply replaceable parts on the assembly line is over. Safety, health insurance, retirement funds, educational and retraining opportunities are considered entitlements. A corporation affects the lives of people by where it places its factory, how and whom it hires, when and how it relocates its plants. It can no longer make such decisions by the fiat of those at the top.

In humanistic or democratic capitalism as De George conceives it, authority flows from the people, from the bottom up and not from the top down. ''Democratic humanism has moved from the political realm to the marketplace . . . the need for consent and democratization in business has increased.'' The lines between business and government, workers and owners, are becoming increasingly blurred. Although workers do not ask to run the factories, they do demand their rights and that their views be considered when the corporation makes decisions; and so do consumers and the general public. Since these demands cannot easily be put in cost accounting terms, the ''social audit'' does not lend itself to precise measurement. Nevertheless profit is not the only goal, and management must be held responsible for satisfying society's demands. If the corporation is to survive in anything like its present form, with ultimate decisions still in the hands of managers, but with adequate responsiveness to the demands of ''humanistic capitalism,'' then major organizational changes will have to be instituted. The last part of De George's essay consists of elaborate recommendations about the specific changes he thinks will have to be made.

Conclusion: The NEXT Step in Business Ethics

De George, by his own account, has come to save capitalism, not destroy it. ''The (proposed) changes taken as a whole fall far short of socialism and workers' self-management.'' ''(Such) changes are the only ones that can enable the corporation to handle the many demands placed upon it and survive in anything like its present form.'' By talking about a new ''model'' or ''image'' of the corporation, and proposing specific structural changes designed to realize that model, De George makes clear what it is that all the essays in this section, to one extent or another, are really driving at: substantial changes in our understanding of how business is to be conducted, but changes that fall short of overthrowing private-investment capitalism and replacing it with something else.

Let us look more closely at what is being advocated. Essentially our authors are not urging that business abandon the set of goals that the conventional ''self-interest'' analysis attributes to them: profit, control of markets, long-range financial stability, and so on. Instead, side by side with these goals, business is being urged to adopt the goals of ''social accounting'': fundamental morality, fairness to all affected, humane treatment of employees, resolution of specific social problems, and the like. Instead of having a single set of goals (not necessarily fully compatible, but all unified as being in the interest of the organization itself, and therefore presumably its managers and stockholders), business is being urged to adopt a double set of goals: self-interest *and* social responsibility. Business under the conventional conception employs all its resources on the former set of goals; under the new proposals, presumably a certain proportion of business resources will be employed to achieve the second set.

How will this affect patterns of investment? If we assume that investors invest in order to maximize return, a business that adopts social accounting will find itself at a disadvantage, compared to businesses that operate in the usual way, in raising capital. If the business devotes, let us say, 20% of its resources to social goals, investors will be required to accept a return on only 80% of their investment, and hence will be disinclined to invest in such an enterprise, provided they have an alternative opportunity to invest in another firm, equally efficient in using resources, in which 100% of the resources are devoted to return on investment. It is conceivable that the socially responsible firm will have great difficulty raising capital and might go out of business altogether. To prevent this, one of two things will have to be done. Either the government will have to support the socially responsible firm directly, or opportunities for alternative, "nonresponsible" investment will have to be eliminated, possibly by laws that require *all* firms to devote specified percentages of resources to social purposes. (Since foreign firms would not be subject to such laws, we would also have to prohibit American investors from investing abroad.) The point is that voluntary compliance in the context of private investment would seem to be unworkable.

Of course this is overstated. Currently there is considerable voluntary compliance with ethical norms, and to suggest otherwise is to misread the ethical climate of business. A *pure* tough capitalistic ethic—in which firms are allowed to commit murder, arson, and theft if they can further their own long-range interest by so doing—is not held by anyone (at least publically) within or without the business world. Corporations currently devote resources to the support of universities, public broadcasting, community organizations, and other good works where the benefit to the corporation (via enhanced image) is unmeasurable. No doubt the extent of voluntary compliance with ethical norms could be much greater than it is. Nevertheless, it seems reasonable to suppose that any extensive shift from self-interest investment to social accounting will not be possible without a significant increase in government control. So long as a firm can save money by polluting, for example, it will be impossible for it *not* to pollute, in order to avoid being placed at a competitive disadvantage with respect to its polluting competitors. Since firms presumptively cannot (because of antitrust) simply agree to stop polluting, thus raising everyone's costs (and prices to customers), and since they probably would not agree even if they legally could, governmental regulation that imposes costs on all polluting firms is inevitable if pollution is to be abated. (Government could in addition pay the cost of required antipollution devices. Essentially, the question is whether taxpayers or customers should pay, given that a certain level of clean air and water has been mandated.)

This is not the place to debate the desirability of even greater government control of business. There is no reason not to believe, however, that social responsibility of the kind envisioned by the essays in this section, could not be attained, should we want it, and were we willing to accept the greater role. It is quite another question, however, whether we *should* want such enhanced social and moral responsibility to be a major goal of business. It is not an answer to hold that a business is essentially no different from an individual, and is hence subject to exactly the same rules of morality (ordinary morality) that govern individuals. No doubt De George is strictly correct in holding that, in the final analysis, corporations are strictly non-entities, and dissolvable into the individual managers who make corporate decisions. But the question still remains whether managers who make corporate decisions should consider themselves bound by the same rules they would be bound by in their capacity as private actors. Nor does it further discussion by pointing out that even the "tough-minded" managers concede that they are bound by *some* rules of self-restraint (not to burn down their competitor's factories, e.g.). The question to be debated is whether they are bound by *all* the ordinary rules of self-restraint, or whether their position as managers gives them special responsibilities that exempt them from some of these rules. Should managers worry about the effect on the local community's tax base when they consider whether to move their plant to a more favorable location? When they lay off employees, do they have a responsibility beyond union contracts to those who have given the firm long years of faithful service? Should they provide special training programs for unskilled and hence unemployable minority workers? May they pay bribes to foreign officials in a country where bribery is considered an ordinary cost of doing business? One achievement of the essays in this section is that they make clear that the usual justifications of the "tough-minded" answer to these questions—greater social freedom, efficiency, competitive necessity, responsibility to stockholders, and so on—are open to serious questions. It is now time to take another step in thinking about business ethics. If we assume that, on the whole, the free market efficiently uses resources and allocates costs fairly to ultimate consumers and if as a general rule unethical practices are imposed by the market, then ethics in business represents an interference with this efficiency-optimizing mechanism. (To the extent that unethical practices are the consequence of not having a free market, the answer would seem to be to create a free market, not to demand profit-reducing ethical behavior.) In such a situation, ethical behavior imposes an added cost that must be paid by someone. If businesses are not to pay bribes where bribes are necessary to secure business, American firms must do without some foreign sales, investors without dividends, workers without jobs. If corporations are to pay "severance fees" when they leave town (or else remain in town absorbing higher labor costs), prices to customers must rise to reflect this extra expense. Ethics, like anything else, has its costs as well as its benefits; what is needed as the next step in clear thinking about business ethics is to analyze these costs and benefits and see to it that they are distributed fairly and rationally. This is a step that, regrettably, is beyond the purview of the current volume.

Is Action Within the Law Morally Sufficient in Business?

L. Duane Willard

There appears to be a widely shared belief among people in business that the ideals, principles, and rules of ordinary morality do not apply to them, or apply in a weaker way. "Business is business, morality is morality." "Business is a dog-eat-dog world." "Caveat emptor." Such ideas suggest that various moral convictions in ordinary life have no place, or little place, in business.

Various reasons for this might be found. For example, it might be suggested that because the business person renders goods or services to society that it needs but otherwise would be without, business is justifiably immune from ordinary moral judgment. Or, it might be claimed that the relationships, tasks, and roles of business people are so different from everyday affairs—that they are more complex, require more technical knowledge and skills, and assume a greater commitment in terms of economic expense and work—that these things somehow place the activities of business on a moral plane different from ordinary life.

There is another reason for this belief, which seems quite prevalent among business people, and is not absent from public opinion at large. It is the related belief that if a person does not break any laws, then what that person does is not morally questionable. Or in other terms, it is morally sufficient to act within the law. Let me quote Albert Z. Carr, a business consultant, concerning the view of many in the American business community[1]:

> We live in what is probably the most competitive of the world's civilized societies. Our customs encourage a high degree of aggression in the individual's striving for success. Business is our main area of competition, and it has ritualized into a game of strategy.

> The basic rules of the game have been set up by the government, which attempts to detect and punish business frauds. But as long as the company does not transgress the rules of the game set by law, it has the legal right to shape its strategy without reference to anything but its profits.

This statement clearly puts the emphasis upon legality as the practical, if not theoretical, normative standard of business activities. In the remainder of this paper I shall attempt two things: I shall indicate several notions that apparently lead to the belief that it is morally sufficient to act within the law. Then I shall consider whether these are good enough grounds for believing it is morally sufficient to act within the law.

I

In calling attention to the notions that may lead people to the view that in business it is morally sufficient to act within the law, I do not pretend to be exhaustive. There probably are some matters I have left out, perhaps some quite significant ones. But those that I include will, I hope, prove adequate to make the point.

First, I take it as obvious that many laws are institutionalized expressions of moral attitudes and opinions about certain kinds of actions. Most societies have laws intended to protect people in one way or another against such actions as murder, rape, stealing, armed robbery, breach of contract, and many offenses. These laws are established because it is believed that such actions are morally wrong, ought not to be done. Now given this fact, there can develop the belief that, generally speaking, the laws of the land incorporate the really important, significant moral beliefs and convictions of people; or, if the laws do not yet incorporate them, that they should do so. Of course we may not be able to legislate morality, but we can and do legislate concerning many areas that are believed crucial to the well-being of people's lives, including their business activities. And this may make it seem that action within the law is morally sufficient. Indeed, I suspect that it may seem to many business people that, if they abide by *all* the laws pertaining to their businesses, there is hardly any way they can go wrong morally.

Second, laws are created and enforced in ways that may make more of an impression upon people than do the moral principles and rules operating in our ordinary lives. Specific persons create and abolish laws, specific persons enforce the laws, and specific persons interpret the laws and make decisions regarding punishments and settlements of conflicting interest. But with morality, it is not the same. There are no specific persons duly designated, or even publicly recognized, as appropriately situated to create, maintain, and enforce moral principles and rules. There is no person or group whose proper task it is to assign punishments for violations of moral rules and principles, or to settle moral conflicts of interest, although, as we well know, some persons proclaim, or assume, themselves to be the moral guardians or watchdogs of society. Moreover, the definiteness or specificity of law often gives a somewhat clearer focus to responsibility than does morality. And, the onerous consequences that can come with conviction, or even suspicion, of breaking a law may be heavier to bear than with the breaking of a moral rule or going against conscience. It will not be surprising, then, to find the belief that laws impress people more, have a more direct and extensive influence upon conduct, than does ordinary morality. And people in business, who must deal not only with the laws that affect all of us, but also with the host of other laws that affect their activities as business people, might all the more easily conclude that it is morally sufficient to act within the law.

Third, it is almost, if not completely, trite to be reminded that the moral beliefs and attitudes of people differ widely on various sorts of things—e.g., abortion, capital punishment, homosexuality, and war, to mention but a few. Moreover, even a rather cursory glance at the ethical theories of philosophers past and present can give one the impression that there may be no end to the number of different points of view—egoism, hedonism, act and rule utility, act and rule deontology, contractualism, intuitionism, emotivism, prescriptivism, existentialism, and so on. Because of such practical and theoretical disagreements, some may be led to the belief that the moral consciences of people are somewhat untrustworthy, that ordinary moral convictions are too often the result of raw emotion, or self-interest, or group pressure, or personal taste. And some may claim that even among the experts on moral theory anyone's opinion is as good as anyone else's. But a civilized society could not survive, let alone flourish, were such variable deliverances of conscience allowed to control, and were conflicting philosophical theories to be taken seriously in the actual affairs of life. What is required is order and stability, and these can arise only out of moral agreements that lead to the establishment of laws in the relevant areas of human relationships. And it goes without saying that the activities and structures of business depend greatly upon social order and stability, much of which is made possible by moral agreements incorporated into legal formulae. Thus it is possible to conclude that it is positive law that really counts, rather than the vacillating spectrum of ordinary moral convictions and ethical theories.

My next point has to do with the suggestion that morality can be more restrictive of freedom than is legality. Generally speaking, the moral codes of societies contain prescriptions concerning more kinds of human activities and relationships than do their legal codes. If we were to draw up two lists, one giving all the kinds of actions for which there are legal measures in a given society, the other giving all the kinds of actions for which there could be moral maxims, it is obvious which list would contain the most entries. There appears to be a sense, then, in which morality could be more restrictive of freedom than is legality. Thus, some may be persuaded that to act within the law is enough, that much moral opinion and judgment apart from legal support may be nothing more than unjustifiable interference in the affairs of people. And in the realm of business, especially where free-enterprise is presupposed, we are likely to find the opinion that morality can put such restraints upon business that it would lack the freedom necessary to succeed as business.

Fifth, there is a very complicated and deeply entrenched set of interrelationships between business and government. Through such things as taxation, regulatory agencies, public corporations, and government contracts, the government in part controls business. On the other hand, through lobbying, campaign contributions, economic advisory boards, membership on regulatory agencies, and other forms of influence, business in part controls government. So interrelated are business and government that in the minds of many the two could not exist without each other. "The business of America is business," it is said. And there are those who apparently take this to mean that the business of government, as well as the citizenry, is the promotion and protection of business. It is small wonder, then, that there can arise the belief among businesspeople that the only moral sanctions that really count in business are the legal sanctions that accompany the connections between government and business.

Finally, some evidently think that, whereas certain laws are appropriate and applicable to business, the convictions of ordinary morality are not. In business, where competition is central to a free enterprise system, there may have to be some laws that protect people from certain kinds and degrees of harm and injustice. But a businessperson who goes beyond what the law requires and conducts business affairs by moral convictions puts the business at a competitive disadvantage. First, he or she cannot be sure that the business' competitors will also adhere to moral principles; indeed, one can be fairly sure that many of them will not. Secondly, different businesspeople might adhere to different moral principles, with disastrous results for some. A business cannot survive unless it makes a profit. As one wag has put it: "If your output is more than your intake, then your upkeep is your downfall." If a person attempts to run a business in accordance with, say, the Golden Rule, that business may show little, if any, profit: the competitors may do unto such a person those things that he or she would not do to them, though they might be adhering to certain other moral principles. It will come as no shock, then, that a principle often followed in business is: Do unto others before they do unto you.

Moreover, just how is one supposed to know how to apply ordinary moral principles in business? Suppose someone in business attempts to follow the moral principle that we ought to try to produce the greatest amount of good for the greatest number of people that our actions may affect. Is it at all realistic to believe that a businessperson could actually carry out this principle in ongoing relationships with customers, employees, and competitors? He or she would have to spend so much time and effort trying to figure out whether specific actions and transactions were really resulting in the greatest good for the greatest number that there would be little, if any, time and energy left to properly manage the business. Or, consider more specific moral duties or rules of ordinary life, such as gratitude. Perhaps morally we ought to show gratitude to our parents, teachers, friends, neighbors, and even strangers who do things for our benefit. But what sense does it make to claim that a businessperson morally ought to show gratitude? So, it is claimed oftentimes that the only rules of the business game are those set down by law, *plus* any other rules or strategies that are conducive to profit-making. And these latter rules or strategies are comparable to the rules or strategies of poker, where mistrust, lying, various forms of dishonesty (from the perspective of ordinary morality) are not merely permissable but fundamentally necessary if one is to succeed.

II

At least these six considerations, then, may lead people to the belief that in business action within the law is enough. Perhaps there is some substance to them. But are they conclusive? I shall take the position that they are not, that neither separately nor together do they show that it is morally sufficient for businesspeople merely to act within the law. In the order previously given I shall deal with the first four points rather briefly, and spend a little more time on the last two.

First, certainly it is true that many important moral convictions are made the subject of law. But not all are. For example, many people are profoundly convinced that we ought not to treat others merely as means to our own ends; in other words, it is wrong to simply "use" other people. But to my knowledge there is no specific law the purpose of which is to enforce this conviction. And where are laws that deal with such things as selfishness, rudeness, friendship, sympathy, moderation, and integrity? Yet these are all matters of serious importance in ordinary morality. The suggestion that it is morally sufficient to act within the law is not adequately supported by the claim that all really significant moral convictions are already contained within law, for they are not. Neither in business nor anywhere else will it do to assume that, if we abide by the laws of the land, this completely fulfills our moral responsibilities.

Regarding the notion that laws make a greater impression than do ordinary moral convictions, that seems true in many cases. But not with everyone, and perhaps not even with the majority. At any time we may find people, at times great numbers of them, who are much more impressed and influenced by some moral belief than by law. Some are prepared to commit civil disobedience with regard to various laws and suffer the consequences. Moreover, even if we were to admit that laws can have a more direct influence upon the thinking and conduct of people, it does not follow that these laws are more important than ordinary moral convictions. After all, not only are at least some laws established *because of intensely held moral convictions, but also some laws that make quite an impression are nevertheless repealed because of moral conviction.* Again, it may seem silly to deny that law makes a greater impression than morality, since punishment for breaking laws is often more severe than for violating the moral code. Still there is countervailing evidence. Penalties for violating some laws are not all that heavy to bear, especially if one's business is doing well enough to afford the penalty. People, including some in business, find various ways of circumventing laws. Threats of punishment for crime, even very severe punishment, evidently do not prevent some from committing those crimes. And, some who break the law turn themselves in because, they say at least, their consciences demand it. Not only is it questionable that law makes a greater impression on peoples' conduct generally than does morality, but also, and more significantly, the influence of law upon people can be simply irrelevant to the question of what is morally important. No doubt people in business are much impressed and influenced by all the regulations they confront. But these regulations do not exhaust the moral responsibilities of business people. Indeed, some may believe they have a moral duty to eradicate various of these regulations.

With respect to the variations of moral opinion in ordinary life and the theoretical differences among moral philosophers, nothing is necessarily implied about the importance of ordinary moral convictions compared to law. For example, there is disagreement among people at large and among philosophers about pornography, although pornography has been and is a very profitable business for some. But surely it is not safe to conclude that pornography as a matter of moral concern is of less importance than anything and everything that is included in the laws of the land. Indeed, if anything, serious moral disagreements in ordinary life and among ethical theorists on certain issues suggest that these issues are at least as important as many things that are made subject to law. On the other side, agreement on an issue, and even establishing a law regarding it, is no assurance whatever that the issue is very crucial morally, or any other way. People do concur on, and make laws regarding, some of the most trivial and absurd things imaginable, a fact that no doubt many a businessperson can appreciate.

With regard to freedom, I think first that the number of cases in which morality is actually more restrictive of freedom than legality is quite small compared to the other way around. Moreover, in many instances where morality does appear to be more restrictive, the moral code is often taken as an unwritten legal code. And although there are more kinds of human actions that could be affected by morality than by law, this does not mean there actually is more restriction of freedom by morality. Many, many kinds of actions that are potentially subject to moral appraisal are not actually so treated in most societies. It is not altogether clear, then, that ordinary moral convictions are more restrictive of freedom than is legality—generally speaking. Of course in business there is little doubt that moral convictions and judgment can have a restrictive influence on economic success. But the businessperson cannot argue without begging the question that one's personal freedom ought not to be curtailed by moral considerations that go beyond the law, for that is the very issue at stake. Without getting into the problem of what free-enterprise means, I want to strongly suggest that operating in a free-enterprise system does *not* by definition mean operating free of all moral constraints beyond the law.

Next, we may agree that interrelationships between business and government are very complex but necessary. What follows from this? It does not follow that the only moral sanctions that should apply to business are those contained in the laws arising out of the interplay between government and business. There are some laws arising out of these relations that many people find morally questionable (e.g., tax loopholes for the rich). Moreover, of serious concern is how well government enforces the laws pertaining to business. In perhaps many cases, the failure to enforce the law can only be described as morally scandalous. Of equal, if not more, concern are some of the ways in which business attempts to get laws passed that are favorable to itself. There are, of course, legal sanctions against, for example, bribery; but it happens nonetheless, and in various forms. We might be charitable and say that the practice in business of buying what one wants legally and politically results from professional blindness, brought on by the pervasive practice of buying what one wants in other business transactions. But that only explains the practice, it does not justify it. Perhaps one of the chief problems is that business fails to realize clearly, or chooses to ignore, the fact that government should function for the well-being of all facets of society, and not just for that of the business community. Serving this vast variety of interests is a very difficult task. Still, from the moral perspective the idea of justice requires that government, and the laws it promulgates and enforces, must be as fair as possible to the various elements of the social whole. That laws arising out of the interdependence of business and government are necessary, then, is obvious. What is not obvious is that these laws are morally sufficient for justification of anything one can get away with in business.

Finally, what are we to say about the claim that the ideals, principles and rules of ordinary morality are not appropriately relevant to and applicable in business? Regarding the notion that to conduct business in accord with moral principles would put one at a competitive disadvantage, what must be immediately pointed out is that competitive disadvantage is a relative affair. For being disadvantaged in business means not having the opportunity to make as much profit as someone else similarly situated. And this in turn assumes that one wants or desires at least as much profit as others are making. The issue of competitive disadvantage, then, comes down in large part to what one wants, to how much profit will satisfy. Now so far as I can see, there is no moral problem with wanting to make a profit. In fact, I am not convinced there is anything morally bad with wanting to make all the profit one can. Some would call this greed; and if we are to call it that, then I am saying that greed in and of itself may not be all that morally suspect. But the trouble comes with the frequent consequences of satisfying greed. For in trying to get all we want, we can be, and often are, led into doing harmful, unjust things to other people. But that is not the only harm we do in acting out our greed. We may also harm ourselves in various ways: develop a narrow, inhibited view of life; become slaves to success and money; give ourselves ulcers and heart attacks; destroy our friendships; ruin our happiness with our families; and so on. If the businessperson insists upon satisfying the desire for more and more profit by harming him- or herself and others, then no doubt it can often be done, so long as one does not run afoul of the law. But let us be done with the rationalizing excuse that such a businessperson is not doing anything immoral, that the principles and rules of ordinary morality do not apply in his or her case.

Regarding the claim that it is difficult, or impossible, to know how to apply the principles and rules of morality in the context of business, surely we must ask some questions. Is there good reason to believe that it is much more difficult for a businessperson to apply ordinary moral convictions to his or her activities than for a doctor, a lawyer, a farmer, a teacher, a busdriver, a politician, or a secretary to do so? Is there, after all, such a great difference between being an honest (or dishonest) person in everyday life and being an honest (or dishonest) person in business? The moral perplexities of life are many and profound, but this does not excuse us for ignoring them. The businessperson may find it difficult to know what ought to be done in various situations that accords with ordinary moral principles and rules, but why should we think it is not incumbent upon that individual to try to be as morally honest as other people? None of us possess perfect moral character or knowledge, but how can this be used to justify excluding people in business, or any other field, from the scope of ordinary moral convictions? And here it should be noted that there are some, perhaps many, businesspeople who go beyond the law and try to conduct their affairs to some degree within ordinary moral principles and rules. At least some of them have succeeded, and succeeded quite well. There are those who claim that virtue is its own reward. But where is it written that virtue cannot possibly accompany fat bank accounts? Business people do not have to be dishonest, any more than anyone else does. They, like the rest of us, can choose to be dishonest if they wish. But neither they, nor the rest of us, can legitimately plead that we are not really dishonest because the principles and rules of ordinary morality do not apply to our activities. It is interesting to note that businesspeople are usually not hesitant to tell us they are honest. On the other hand, if they are charged with being dishonest, they may attempt to convince us that the concept of dishonesty does not apply to them. But of course they cannot have it both ways. If honesty can apply to them, so can dishonesty; if dishonesty does not apply to them, then honesty does not apply either.

It is not unreasonable to hold that some moral principles and rules may not be very relevant or practically applicable in business. For example, I am not sure it makes any sense to say a business person has a moral duty as a businessperson to show gratitude to anyone. Nor do I think it obvious that a businessperson does wrong if he or she fails to conduct business affairs with love. But it does not follow that there are *no* ordinary moral convictions that are relevant and applicable in business. We do not consider the person with whom we do business as morally obliged to save for us all the money possibly savable in our transaction. But neither do we consider it morally proper for that businessperson to get all the money remotely obtainable from us by any means possible. The businessperson may not have a clear duty to bring about the greatest good for the greatest number of people, but is that person also morally excused from trying to minimize the amount of harm his or her business actions might cause? As mentioned earlier, there are those who think of lying and deceit in business as comparable to those ploys in a poker game. Ordinary moral convictions, by and large, include the belief that in some cases it is morally permissable, or even obligatory, to lie or deceive. These are mostly cases where lying and deceit will prevent harm or injustice from being done to people. But what is the justification for lying and deceit in the business realm? The bottom line is profit. Of course it might be claimed that if a businessperson does not lie and deceive, then harm may be done to the business by its competitors who will steal its customers by lying and deceit. And so it may be, as it is in poker. But how does this justify lying and deceit as a practice throughout the business world? Even if we try to justify lying and deceit in business or poker on the basis of harm by competitors, this entirely omits from the account the harm and injustice that lying and deceit in business, but not necessarily in poker, can bring upon the public.

So, I conclude that it is not morally sufficient in business merely to act within the law. For the law may not, and often does not, adequately provide sanctions against the harm and injustice that people in business can inflict upon others. Ordinary morality provides sanctions that are missing from law, or are too weakly expressed in and enforced by the law. I am not so naive as to imagine that the convictions of ordinary morality are sufficiently powerful and resourceful to prevent all misconduct in business. Perhaps nothing can do that. But this certainly does not mean that morality is less important than law in serving the causes of justice and the common good. Morally speaking, it has never been enough to be a law-abiding citizen, inside or outside of business, although even that can be a good start.

Reference

[1]Albert Z. Carr, "Is Business Bluffing Ethical," *Harvard Business Review* **40** (January-February 1968).

The Theoretical Twist to Irresponsbility in Business

Robert V. Hannaford

In 1970 the Nobel Laureate economist Milton Friedman wrote an article entitled "The Social Responsibility of Business is to Increase Its Profits." He began his article with this paragraph:

> When I hear businessmen speak eloquently about the "social responsibilities in a free-enterprise system," I am reminded of the wonderful line about the Frenchman who discovered at the age of 70 that he had been speaking prose all his life. The businessmen believe that they are defending free enterprise when they declaim that business is not concerned "merely" with profit but also with promoting desirable "social" ends; that business has a "social conscience" and takes seriously its responsibilities for providing employment, eliminating discrimination, avoiding pollution and whatever else may be the catchwords of the contemporary crop of reformers. In fact, they are—or would be if they or anyone else took them seriously—preaching pure and unadulterated socialism. Businessmen who talk in this way are unwitting puppets of the intellectual forces that have been undermining the basis of a free society these past decades.[1]

Thus, Friedman lends his personal prestige to a deliberate abdication of responsibility. He and other economists have invoked the authority and intellectual defense of economic theory to support unethical attitudes and to give a view of the role of the corporation in society that is oversimplified and confusing. These points become clearer as his article progresses, but begin to emerge in the opening paragraph, in his rejection of any notion of special social responsibility for business, in his use of scare quotes, and in his comparison of the businessperson who would be responsible with Moliere's M. Jourdain, who would learn to speak prose. We are asked to share the laugh on the businessperson because we are supposed to agree that he or she, as one who succeeded in business, could no more miss being socially responsible than M. Jourdain could fail to speak prose in his normal activity. References to the social responsibilities of business are put in scare quotes to indicate that we are to avoid using these supposedly mistaken expressions and the word "merely" in the clause 'business is not concerned "merely" with profit' is put in scare quotes to indicate that in the pursuit of profit business already bears all the social responsibility it should be concerned with.

Friedman is not alone in taking such a stance. But I want to show how he is able to foster such unethical attitudes by giving a twist to what is already a vague and ambiguous element in classical economic theory. Once that point is made, it will be clear that Friedman's own account of responsibility is incoherent and, if we follow our standard conceptions of responsibility, we will want to say that businesspeople and particularly those who are executives and directors of large, publicly owned corporations, do have special social responsibilities. Moreover, they are responsibilities whose excercise in no way threatens a free market.

I. Classical Ambiguity and the Friedman Twist

Classical theory's explanation of the free market turns on the explanation of the motive for economic behavior, an explanation based on the phrases "bettering one's condition" and "pursuing one's own interest" or on some variation of those phrases. The Friedman twist is possible because of the vagueness and ambiguity of this phrase (or its substitutes). The original formulation facilitates, though it does not require, Friedman's view. In Smith's *Wealth of Nations,* he wrote that the basic source of economic activity is found in our individual "desire of *bettering our condition,*" a desire that " . . . . comes with us from the womb and never leaves us 'til we go to the grave."

He wrote that "every individual is continuously exerting himself to find out the most advantageous employment for whatever capital he can command." This assumption (whether literally true or not) is central to the explanation of the free market; the drive it describes is the fundamental force of the price mechanism. From this assumption we can argue that capitalists who seek what is most advantageous to them (in the form of profits) will become informed about, and therefore produce, what will fetch top price, thereby satisfying the greatest economic demand. Similarly, laborers seeking top price for their services will move to the place where the demand for them is greater. Thus, Smith held, each seeking personal financial advantage is "led by an invisible hand" to contribute to the general welfare. Smith's account makes plain how those who do pursue their own financial gain might thereby contribute something valued by the rest of the community. And, thus, their contribution might be regarded as fulfilling some part of their social responsibility. However, nothing is said nor can be inferred to the effect that this contribution fulfills the *sum* of one's social responsibilities. That is Friedman's twist. However, the looseness of the phrase has been exploited by other economists to twist the theory to the same effect.

The notion of each of us striving to better his or her condition has the same looseness of meaning that the notion of pursuing one's interest has. Some senses of that loose expression are true: no doubt we would agree that each person seeks to better his or her condition *as he or she perceives* what would better it, and that each of us uses personal capital in what is perceived to be the most advantageous fashion. However, it is not clear whether our interest or condition is to be taken to include or exclude the interest of others; it is not clear whether our interest is to be given a purely financial interpretation. If the phrase is taken in its broadest, loosest sense, it says that each person is concerned with his or her own affairs and tries to improve on them to the greatest possible extent. That is a truism, but it has the fatal defect that it will not enable us to predict what different people will take to be included, under "bettering their condition" and so as a consequence, it would not enable us to predict or explain any specific kinds of economic behavior. On the other hand, if we give it a stricter interpretation, so that it has some explanatory significance, and say that each person is continuously seeking to improve his or her financial condition, it turns out to be false. It is not true that we always so seek, or ought to seek, the economically most advantageous use of our capital. Not all do: not the ascetics, not the improvident, not the other-wordly, not those who stuff money in mattresses. Nor should we always: not when our friend is starving, not when our neighbor needs medicine only we can provide. Economists when confronted with this problem have sometimes sought refuge in the notion that all the things a person seeks may be regarded as having a kind of utility; they thus seek to rescue the central selfishness axiom. That notion is true, but the rescue is made at the same cost. To say that all persons will seek what they regard as useful does not tell us anything about what in particular they so regard, nor what they will seek. So some economists have done a kind of shuffling two-step back and forth between the different interpretations of the selfishness axiom: first they say, take it as vague and general (in order to have it accepted as true); then, having established that people *seek their own interest,* they assert: We have a right to say that each person seeks his or her own financial interest continuously. When someone raises a question about that proposition,[2] they say: "But surely you don't mean to say that you think people sometimes don't seek what's in their interest?" Thus they hide behind the broad and vacuous sense of the axiom. What this shuffling on the self-interest axiom has done is to provide a defense of egoism: it tells us that *everyone* is selfish, and, if you will but leave them alone in their selfish pursuits, they will contribute their bit to the economy and we will all be happier and more productive. *Greed* is good for us: it makes everybody happy. *Only confused* do-gooders worry about social responsibilities.

Smith's point, of course, was that, whatever people were interested in, they could obtain more easily and produce more efficiently in a free market. He showed how a free market was superior to government control of what was bought and produced. A free market will provide the greatest liberty of choice for the buyer and freedom of work for the producer and the most productive use of whatever capital we have at our disposal.

However, that liberty of choice will still be available even to those socially responsible producers who do not want to get absolutely the greatest financial return on their investment. Suppose that, because of some social responsibility that one recognizes, one is prepared to settle for a lesser return. Let us say that one will thereby avoid polluting one's neighbor's water supply and that one elects to do so. Now clearly this is a move that producers have made and will continue to make without any kind of social control forcing them to do so. Farmers, for example (who compete for their income and are part of a free market), have often taken the extra care and time required to avoid polluting others' water supplies. While doing so, they have spent time and capital that might have been spent in earning a buck. In doing so, they have acted freely, facing a social responsibility other than that provided by their role in the price mechanism. But such a decision does not undermine the freedom in the society or the freedom of the market. To suggest that it will is to give a perverted picture of the price mechanism: farmers can continue to produce what others will want and can make a living; scarcity will continue to increase prices; individuals will still be able to make their own economic choices. So it does not make sense to say (as Friedman does) that such behavior is "undermining the basis of a free society." Liberty is not tied to selfishness. People have remarked upon our responsibility to others since the beginning of time. Those who have remarked on such responsibility can hardly be characterized as "preaching pure and unadulterated socialism"—as Friedman would have it. Such people have seen something that they could do to promote a common good and they have chosen to do it. They have thereby enhanced the stability of free societies. I will later suggest that parallel considerations apply to corporate actions and to the responsibility of corporate officers.

However, of the responsibilities of a corporate executive, Friedman writes:

> In a free-enterprise, private property system, a corporate executive is an employee of the owners of the business. He has direct responsibility to his employers. That responsibility is to conduct the business in accord with their desires which generally will be to make as much money as possible while conforming to the basic rules of the society, both those embodied in law and those embodied in ethical custom.
>
> . . . the key point is that . . . the manager is the agent of the individuals who own the corporation . . . and his primary responsibility is to them.

No doubt an executive officer has a responsibility to make money on an owner's investments. But, in the context, the only point of emphasizing it is to deny social responsibility, to suggest that no consideration of public obligation ordinarily would or should modify a business owner's pursuit of private gain. The chief executive officer is to exploit all financial opportunities available within law and custom. For Friedman, corporate responsibility is vested solely in the hands of individual stockholders. By this account the corporate manager has no special social responsibility that accrues as a result of his or her corporate position; the manager's corporate responsibility (unless the manager is specially charged by the stockholders) is simply to make money for them. Neither economic theory nor the facts of corporate ownership will support such an analysis, though Friedman intimates that they do. He uses this intellectually confusing analysis to promote the bottom line mentality that has become a recurrent feature of contemporary business.

It was the bottom line mentality that lay behind the McDonnell-Douglas executive's decision to rush the DC 10 into production to get ahead of its competitors, even though their own engineers had warned the management about the poor design of the cargo door latch. It was this inadequacy that resulted in the disastrous crash of a DC 10 at the

Paris airport in 1975.[3] Nor is this an isolated case. Others have noted that

> . . . management's tendency to measure performance on the basis of short-term, objective standards often leads a firm to ignore the social impact of its operations, which is often difficult to measure. Environmental pollution most readily comes to mind. The corporate decision maker's interest in achieving results that promote his personal goals under the existing reward system often causes the corporation to generate more harmful pollution in order to reduce measurable costs and increase measurable profits.[4]

Studies have shown that where large firms place great emphasis on making the most money, their chief executive officers are ''single-mindedly, almost slavishly, committed to achieving'' a showing of maximum short-term profits.[5] When Roderick M. Hills was chairman of the Securities and Exchange Commission, he noted the temptation of management to be irresponsible:

> When reported profits decline to such an extent as to threaten the serenity of their well-paid isolation, some managers are tempted to change the accounting, the figures or the morals of their company in order to present a more pleasing profit picture.[6]

II. Corporate Officers Decide—Rarely Stockholders

It is apparent that Friedman's denial of our social responsibilities turns on the sense of the term ''responsible.'' For our discussion there are two standard relevant senses: one of the senses is that of holding someone responsible who is identified as the agent or cause of some change, a sense found in such questions as ''Who's responsible for the broken window?'' Another sense is found in our question whether someone is able to respond to a situation in a way that is in accord with the values of others. It is a matter of knowledge and ability and it is used when we ask whether an 18-year-old is mature enough to be responsible. These usages evolved from describing persons and their officers as well. Corporations bring about changes that no single individual could accomplish and corporate executives have power and can recognize possibilities for change that others cannot. As Edward Mason observes:

> We are all aware that we live not only in a corporate society but a society of large corporations. The management—that is the control—of these corporations is in the hands of, at most, a few thousand men. Who selected these men, if not to rule over us, at least to exercise vast authority . . .[7]

If we are to give an account of contemporary life, we must be able to speak of the special responsibilities of the corporation and the corporate executive.

However, this will be a suspicious move, as far as Friedman is concerned. In the article cited above, he suggests that one is linguistically confused if one speaks of corporate responsibility and of the responsibility of the corporate executive.

> The discussions of the "social responsibilities of business" are notable for their analytical looseness and lack of rigor. What does it mean to say that "business" has responsibilities? Only persons can have responsibilities. A corporation is an artificial person and in this sense may have artificial responsibilites, but "business" as a whole cannot be said to have responsibilities, even in this vague sense.

One should observe that "the responsibilities of business" refers to the responsibilities of *those in* business and note that it is standard to speak of groups that perform a particular job as having a kind of social responsibility because of that job. We speak of education (those in education) as having a kind of responsibility, of (those in) government as having certain responsibilities in a matter, and so on. There is no linguistic confusion here. And there need be no linguistic confusion in speaking of those in business of having special responsibilities that accrue to them as a result of their role in business. Those in business have special responsibility not to misrepresent their product and to see to it that their product does not present any hidden dangers to the consumer. If we deny that businesspeople have special responsibilities, it will not be because our language does not permit it.

However, Friedman also holds that it is not legitimate to speak of one's responsibilities as a corporate executive, except as one's responsibilities to one's stockholders.

> What does it mean to say that the corporate executive has a "social responsibility" in his capacity as businessman? If this statement is not pure rhetoric, it must mean that he is to act in some way that is not in the interest of his employers. For example, that he is to refrain from increasing the price of the product in order to contribute to the social objective of preventing inflation, even though a price increase would be in the best interests of the corporation. Or that he is to make expenditures on reducing pollution beyond that amount that is in the best interests of the corporation . . . .

We note that the argument and Friedman's objections turn on the vague and ambiguous phrase "in the best interest" of the corporation, the phrase that dogs economists' explanations. Here we do not know whether one is using the term to speak of short-term profits, the firm's long-term stability and financial security, prestige in the community, or what. Of course, it will make a material difference which of these senses is being considered by the chief executive officer when choosing what is in the firm's best interest. However, if the officer considers long-term financial interests, stability, or prestige in the community, that officer might well want to consider what might be called the social responsibilities of the firm. Certainly a great many corporations spend a lot of time and money trying to convince us that they are socially responsible. Yet in cases where they are, Friedman finds that:

> . . . the corporate executive would be spending someone else's money for a general social interest. Insofar as his actions in accord with his "social responsibility" (always in scare quotes) reduce returns to stockholders, he is spending their money. Insofar as he is raising the price to customers, he is spending the customer's money.

For Friedman the executive's responsibility is to those hiring him or her, and thus the executive has no right to make decisions in their behalf.

> The stockholders or the customers . . . could separately spend their own money on the particular actions if they wished to do so. The executive is exercising a distinct "social responsibility" rather than serving as an agent of the stockholders or customers . . . only if he spends the money in a different way than they would have spent it.
>
> But if he does this, he is in effect imposing taxes, on the one hand, and deciding how the tax proceeds shall be spent on the other.

One fulfills one's responsibility as an agent only if one spends the constituents' money in the same way that they would have spent it. But the phrase "that they would have spent it" is not clear. Does it refer to how they would have spent it, given their present knowledge and consequent desires? Or does it refer to how they would have spent it if they were given the knowledge and know-how of one exercising the corporate position? Is it to be supposed that all stockholders and customers would have reached the same conclusion? Would that be before or after they had acquired the understanding?

It may be suggested that we are to have the corporate executive act as the agent of the desires that we as stockholders now have. It may be suggested that all customers will want the commodity at the cheapest possible price and that all stockholders will want the greatest possible return on their investment. But here we run into a difficulty of deciding what we mean by "possible," in "cheapest possible" and "greatest possible return." Possible with what else? We need to know for how long a time period it is to be possible, for what cost to the health of ourselves and others, for what cost to the security of the business and much more. "Greatest possible" and "cheapest possible" will mean very different things depending on which of these many specific conditions is to be preserved. An executive of a publicly owned corporation cannot know what conditions will be considered important for all of the corporation's stockholders, nor for what additional reasons. As a holder of stock in California Edison I might say that its corporate executive acted as the agent of my desires when that executive decided to keep a nuclear power plant in operation because I would thereby be able to obtain an 18% return on my investment for the next quarter. But my desire for an 18% return would change if I discovered that the plant was unsafe and that its continued operation posed a threat similar to that of Three-Mile Island or *The China Syndrome*.

Increased information about such consequences will not bring unanimity of desires on the part of the stockholders. It may be that my aging, misanthropic Aunt Ella, now living in London and holding a large block of California Edison stock, will not be concerned to protect either the health of the Los Angeles residents or the long-term viability of the company.

Friedman attempts to answer the question of what conditions are included under the phrase "greatest possible" when he writes:

> . . . (a corporate executive's) responsibility is to conduct the business in accordance with (the owners') desires, which generally will be to make as much money as possible while conforming to the basic rules of the society, both those embodied in law and those embodied in ethical custom.

His answer, then, would be "as much as we can while conforming to the law and ethical custom."

But the answer is unsatisfactory. We cannot appeal to custom, for the judgment of ethical custom about what our corporate responsibilities may be is what we have been unclear about all along. We cannot settle a *dispute* about what our responsibilities are by saying that our responsibilities are what we think them to be. If the nature of responsibilities in individual cases is widely disputed, then ethical custom in such cases cannot resolve the dispute. Nor can we define our responsibility as doing what the law will allow. Fulfilling one's responsibilities involves doing more than avoiding those acts for which one can be legally punished. Harold Williams, chairman of the SEC, remarked that the country's level of social stability and strength is to be measured by the difference between the operative ethical principles and the legal sanctions requiring conformity. The greater the individual responsibility taken without threat of legal sanction, the greater its stability and strength. Friedman cannot preserve a free society by defining our social responsibilities in terms of what the law will allow. All of this disagreement illustrates the possibilities for stockholder disagreement: what they want their company to do will depend in part on what they think it ought to do.

One is led to the conclusion that stockholders in different circumstances will ordinarily differ among themselves as to what policy they think the firm should follow. It will not be possible for the executive officer of a large corporation to avoid spending money (to use Friedman's phrase) in a different way from how they would have spent it. The executive must spend as he or she sees fit.

It does not make sense to speak of the corporate executive as the agent of the stockholders' desires. There could be no construction of Friedman's phrase "how they would have spent the money" according to which large numbers of people of different interests, education, and ability would come to understanding and agreement on a policy to be pursued. Hence, Friedman's notion of the responsibility of a corporation executive as agent of stockholders' desires is unworkable for officers of large, publicly owned corporations.

III. Powers Entail Responsibilities

Corporate decisions must be primarily decisions of corporate officers, not of stockholders. Large, publicly owned corporations have possibilities and resources of such complexity that the stockholders cannot understand them unless they are also involved in management. The Ford Motor Company, e.g., in 1976 had capital assets of over 15 billion dollars and employed 450,000 people. Other corporations have similarly formidable resources, but each is different.

The distinctive features of contemporary society have been created through these corporations. Without them our technology for communication, transportation, and manufacturing could never have come into being. They have transformed our society root and branch. When we give an account of social changes they must be figured in.

These corporations have a kind of life and power of their own, apart from the specific persons who provide them with capital or serve as their officers over any particular time period. But those occupying executive and directorial positions at any time are agents of sweeping social changes affecting us all. For such changes they must be held accountable, if anyone is accountable for anything; such power and capacity carry social responsibilities with them.

Acting to fulfill social obligations is not contrary to a firm's business interests. Business interests include community interests. Corporate decisions should center on what is in the total interest of the corporation. If a corporate decision is well-made, it will be seen to include concern for the community that makes the corporation possible.

If farmers and small businesses can decide to increase their operating costs in order to satisfy social obligations and not thereby threaten their livelihood or the free market, so can publicly owned corporations. The decision process is different in the two cases, but in both cases it turns on the decision-maker's concern for others and ability to do what the common good requires. If either lacks such concern, there is little likelihood that their decision will be responsible and in each case it must be a decision that the decision-maker sees as fit in order for it to be made at all. Corporate decision-makers often have superhuman resources at their disposal for seeing what *is* fit: lawyers, engineers, management consultants, members of the board, and perhaps more. So the decision, often enough, is the result of a sustained inquiry and discussion by the staff rather than the reflection and internal dialog of a single person. By conducting such an inquiry and discussion, the corporate officer does not thereby so increase the cost of production that the firm is removed from competition in the market. That this is so is attested to by the fact the testing and research divisions are typically used for such purposes and are standard parts of large corporations.

Ordinarily we assign responsibility in accord with one's capacity to understand the implications of his or her actions and to control their outcomes. As Bishop Butler remarks, we hold people responsible for actions to the degree that they are able to foresee and control their consequences. If so, business must not only acknowledge its social responsibility, it must often bear an extra burden of responsibility. Each of us feels more prepared to bear a personal share of responsibility to the extent that others bear theirs. Whoever denies these responsibilities thereby denies any respect for others, and whenever such respect for others is extensively denied, individual rights and freedoms are threatened. Businesspeople who have extolled the single-minded pursuit of profit as the Free-Enterprise Way have engaged in such a denial of respect. The economists who have twisted classical theory to defend such egoism have aggravated their threat to a free society.

Notes and References

[1]Published in *The New York Times Magazine,* Sept. 13, 1970. All references to Friedman's work will be to this article.

[2]If the self-interest axiom is to be of any service for explaining and predicting the operation of the price mechanism, we must take it to refer to what is economically advantageous. Yet it is false that ''every individual is continuously exerting himself to find out the (financially) most advantageous employment for whatever capital he has at his command.'' This makes an assumption about our basic value commitments that is false for many. However, if we are describing economic activity in a materially oriented society, the axiom may be rescued as a basis for making statistical predictions about what *classes* of people will do and about general market trends.

[3]Reported by Joseph Iseman, a lawyer for one of the victims of the crash (in a lecture given at Ripon College during the spring 1979 term).

[4]From J. Bower's article on ''The Amoral Organization'' in *The Corporate Society,* Robin Marris, ed. (London: Macmillan, 1974).

[5]From C. Argyris's article on ''The CEO's Behavior: Key to Organizational Development,'' *Harvard Business Review* (March-April 1973), 55.

[6]Quoted by C. Weiss and B. Schwartz (p. 77) in their ''Disclosure Activates Directors'' in *Reweaving the Corporate Veil,* a study in Duke University School of Law's series, Law and Contemporary Society. (1977)

[7]From his Introduction to *The Corporation in Modern Society,* ed. E. Mason (Cambridge, MA: Harvard Univ. Press, 1959).

A Laissez Faire Approach to Business Ethics

Donald R. Koehn

In this paper I examine an ethic that is seldom explicitly recognized as an ethic. It is familiar to many people in business and to faculty and students in the fields of Business and of Economics. Though seldom recognized or articulated as an ethic, the view to be examined is today most widely associated with the *laissez faire* philosophy of Milton Friedman. Twenty years ago, the view was probably most widely associated with the writings of Ayn Rand. Perhaps, as many people believe, the ethic is to be traced back to Adam Smith's *The Wealth of Nations,* where, in a frequently quoted passage, he says, "It is not from the benevolence of the butcher, the brewer, or the baker, that we expect our dinner, but from their regard to their own interest."[1] Those familiar with this passage or with the *laissez faire* advocacy of Friedman or of Rand will recognize an idea that provides the foundation of a business ethic: in a *laissez faire* system one individual's pursuit of self-interest does not have to be tempered by concern for the interests of other individuals because the free markets of such a system integrate the diverse interests of different individuals. The butcher, for example, may be ever so selfish, but is forced by the market to meet the demands of customers in order to fulfill self-interests. This idea, indeed, this ideal is readily recognizable by many businesspeople and people with an academic background of business or economics. Few, however, would be able to identify a writing where the ideal is briefly articulated and vigorously defended as an ethic of business. Fortunately, there is such a writing.

When Jack Hirshleifer, now Professor of Economics of the University of Los Angeles, was Co-editor of the *Journal of Business* he published a paper, "Capitalist Ethics—Tough or Soft?"[2] In the paper

Hirshleifer defended the tough capitalist ethic. Once that ethic and its defense are laid out, readers familiar with "classical" economic theory will discern an ideology that, whatever its origin and whoever its spokespersons, has today a vocal band of champions, who, with the election of Ronald Reagan, have become quite influential. Furthermore, to the extent that readers familiar with the classical theory believe that there is a special ethics of business activities that is less stringent than an ethic appropriate to non-economic matters, they are likely to find in Hirshleifer's paper their own frequently unformulated rationale for this belief. Using Hirshleifer's paper as my stalking horse, I shall argue that the tough capitalist ethic is not an adequate ethic of economic affairs and that the *laissez faire* ideology does not provide an adequate rationale for the position that there is a special, less stringent ethic of business.

Before I make those arguments I shall in the first section lay out Hirshleifer's *laissez faire* position, the difference between the tough and the soft capitalist ethics, and the rationale for there being a special ethic of economic affairs. In sections II and III I shall critically examine to what extent the positions of Rand and/or Friedman agree with that of Hirshleifer. (Readers not interested in Rand or in Friedman may conveniently skip either of these sections.) In section IV I shall indicate why *laissez faire* capitalism justifiably has a strong appeal to many people. My arguments against Hirshleifer's tough capitalist ethic and against his rationale for there being a special ethic of business will be made in section V. I shall then provide some concluding remarks.

Tough Capitalism, Soft Capitalism, and the Ethic of Each

Hirshleifer presents and defends the ethic of tough capitalism polemically. He contrasts it to the ethic of soft capitalism, which he takes to be well-represented by James C. Worthy, then Vice-President of Sears Roebuck and Co.[3]

The soft view sees the business system as convicted of encouraging selfishness and of failing to serve humanity. Businesspeople, misled by the *laissez faire* economic ideology, talk as though they can base their conduct solely upon self-interest. The public can hardly be expected to respect or trust the business system when its spokespersons deny that a businessperson has any responsibility except to him- or herself. Furthermore, although businesspeople are not nearly so selfish as they talk, their conduct is more selfish than it would otherwise be because their actions are influenced by the *laissez faire* rationale. No wonder the public demands the dreaded, stifling laws and regulations of government to impede the freedom of individual businesspersons.

Businesspeople, still according to the soft capitalistic view, need to renounce the discredited *laissez faire* ideology and recognize that they must exercise self-restraint. When one acts in an economic role, a person is to consider not only her or his own interests, but he or she also is to consider the interests of employees, of customers, of stockholders, indeed, of anyone who is likely to be affected by a decision under consideration. One's actions in economic matters must be guided not only by factors relevant to economic self-interest, but also by the ideal of fair play when the interests of all of those to be affected by a proposed action are taken into account. Self-restraint must be at the heart of business ethics, and besides, if people in business activities and relations do not exercise self-restraint, they had better be prepared for legal restraints.

All of this, Hirshleifer insists, is terribly, terribly mistaken. In the first place, the soft capitalistic ethic fails to recognize a fundamental fact about humans. Humans are selfish and we had better take that fact into account. (Some statements of Hirshleifer's article imply that all people are completely selfish; others imply that not all people are completely selfish or even at all selfish. He does consistently insist that social conditions or institutions, particularly those of capitalism, do not make people selfish.) In any case, according to Hirshleifer, Worthy is just dead wrong in implying that in economic activities people are not selfish. Contrary to the soft view, businesspeople mask selfishness with unselfish talk. The ethic of tough capitalism does not try to persuade people in their economic activities to be considerate of others, which people, or most people, cannot do anyway.

The second way in which the soft capitalistic view is terribly mistaken is that it fails to understand the nature of capitalism. According to that view the only restraints upon selfishness are either legal restraints or self-restraint. But at the heart of *laissez faire* capitalism is a third alternative, namely, the restraints of free markets. The person who wants to buy or to sell is restrained by the market competition of others who want to buy or to sell. I may not care at all for the interests of others and government may leave me completely free to do what I want to increase my wealth, but my freedom of action is circumscribed by what people are willing to buy and by the competition of those willing to sell. Furthermore, the genius of capitalism is that its market restraints direct the effects of completely selfish actions into the service of others. Whether as a buyer or seller, I must meet the demand of what other people want as expressed by their willingness to pay. Part of the greatness of *laissez faire* capitalism is that it harnesses the great energy of uninhibitedly selfish behavior to the service of others.

The third way in which soft capitalism is terribly mistaken is that it implicitly but fully accepts the "new managerialism".[4] According to this theory of corporate management, executives of corporations should serve the interests of all people who are affected by managerial decisions. The seeking of profits for corporate stockholders is only one of competing interests that the corporate manager is to serve. The ethic of soft capitalism implicitly requires the acceptance of this divided responsibility on the part of corporate officers. But, Hirshleifer replies, no one can serve two competing masters; at best, a person can mediate between them. The executive is in fact responsible to neither. To hold a managerial group responsible to company employees, to customers, to the public at large, and to stockholders is simply to invite the managerial group to be responsible only to itself. The effect of an attempt of soft capitalism to avoid selfish behavior on the part of corporate managers would be to encourage their selfish behavior. It is far better for all concerned for the self-interest of the manager to depend solely upon the interests of one group, the profits of the stockholders.

For these three reasons, soft capitalism with its ethic must be rejected and we must accept tough (*laissez faire*) capitalism with its ethic.[5]

Hirshleifer does not explicitly state any principles of the tough capitalist ethic or just how that ethic differs from the ethic of soft capitalism. Indeed, he does not discuss an ethic that he clearly identifies as being the tough ethic and his reader is left to infer what that ethic is from his critical discussion of the soft capitalist ethic and his approving discussion of tough, i.e., *laissez faire,* capitalism. Nevertheless, the difference between the two ethics and the principle of the tough ethic can be rather securely inferred by constructing the different ways by which a soft capitalist and a tough capitalist would make an economic decision.

In making a business decision according to the soft capitalist ethic, a person considers (1) the conditions of the pursuit of self-interest, (2) presumably business management principles or principles of prudence that are applicable to the conditions, and always, (3) sometime before a decision is made, the way in which a proposed course of action or policy would affect the interests of all those persons who might be affected by the proposed action or policy. Although the exact formula for balancing self-interests and interests of others is not specified, that one is to be guided by some ideal of fair play for balancing self-interests and the interests of others is clear. According to the ethic of soft capitalism, neither a private party nor a car salesperson would be morally permitted knowingly to sell a used auto with a faltering oil pump to a person who needs to get from Chicago to San Francisco without warning the prospective buyer of the bad pump. Certainly the soft capitalistic ethic would not permit the selling of a used auto with a cracked brake line without warning a prospective buyer of the danger. The interests of other motorists as well as the interest of the prospective buyer would morally compel self-restraint on the part of the prospective seller. Due consideration for others exercises a veto upon the pursuit of self-interests in business activities no less than in personal relations.

According to the ethic of tough capitalism the decision procedure is to omit the third step. In its place one seeks only one's own interests without balancing them against the interests of others. One considers the interests of others only to the extent that enhancing or harming them may affect one's own interests. For example, one might refrain from harming another by selling a defective auto if doing so would be expected to harm the seller's future ability to sell upon the used car market. Nevertheless, the fact that others may be harmed or advantaged is not itself an ethical factor for making an economic decision. The ethic of tough capitalism, then, unlike the ethic of soft capitalism, does not require that one exercise self-restraint for the sake of the interests of others. This much is quite clearly to be inferred from the text of "Capitalist Ethics—Tough or Soft?"

What is not clear is whether the tough ethic requires or only permits selfish economic behavior. Hirshleifer's strong attack upon Worthy's ideal of self-restraint in business and his discussions of selfishness strongly suggest that the principle of the tough ethic is that in economic activities one *ought* to act selfishly. There is a passage, however, in which Hirshleifer says that under tough capitalism "it is not true that everyone *must* be selfish . . . ."[6] To deal with the ambiguity, the principle of the tough capitalist ethic needs to be stated ambiguously as follows: *In economic affairs, one ought, or is permitted, to act selfishly*. That one is ethically required to act without regard for the interests of others we shall call the "strong version" of the tough capitalist ethic and that one is only ethically permitted to act selfishly we shall call its "weak version." In either case, the great difference between the tough and the soft capitalist ethics is that the latter ethically requires self-restraint upon the pursuit of self-interests in the light of some kind of fairness to others' interests, while the former does not require any kind of restraint upon the pursuit of one's own interests for the sake of others' interests.

We can now ascertain how Hirshleifer supports the thesis that there is a special business ethic that is less stringent than an ethic adequate for non-economic affairs. One apparent justification, which we shall immediately dismiss, concerns Hirshleifer's claims that humans are selfish. The view that everyone in fact is selfish is called "psychological egoism," and it is sometimes used to support the view called "ethical egoism," which is the position that everyone ought (ethically) to act selfishly. If humans are psychologically incapable of acting with due regard for the interests of others, it hardly makes sense to say that humans are morally obligated to do so. This kind of justification of ethical egoism by psychological egoism would justify an ethic of selfishness for all human life and not an ethic distinctive of economic matters. But Hirshleifer is rather clear that although the tough capitalist ethic is applicable only to economic affairs, human selfishness extends beyond them. Hence, we need not further consider psychological egoism as a justification of a special business ethic. The important reason for the thesis that there is such an ethic is that under *laissez faire* capitalism free markets direct the effects of selfish actions to the satisfaction of the interests of others. As each person vigorously pursues self-interests, the market allocates benefits to all others who are affected by an action and consequently everyone ought or is permitted to be completely selfish. Accordingly, in economic affairs the free markets of tough capitalism perform the function of self-restraint in non-economic affairs and there is a special, less stringent business ethic.

Having now grasped Hirshleifer's *laissez faire* approach to business ethics, we need to evaluate it. If we could know, however, the extent to which more popular spokespersons of that kind of approach are similar to Hirshleifer's, we could know to what extent our evaluation of his approach is applicable to them. Accordingly, let us briefly consider the positions of Ayn Rand and of Milton Friedman before we evaluate Hirshleifer's position. That evaluation will not depend upon the discussions—which will be criticisms—of Rand and Friedman, and readers uninterested in their positions may omit the next two sections.

Rand's Inconsistent Ethic

The author of the novels *Atlas Shrugged* and *The Fountainhead* explicitly recognizes that she holds an ethic in her book, *The Virtue of Selfishness: A New Concept of Egoism.*[7] Ayn Rand nicely summarizes her ethic in the first essay of the book, "The Objectivist Ethics." The ethical principle of selfishness is there stated:

> An *ultimate* value is that final goal or end to which all lesser goals are the means—and it sets the standard by which all lesser goals are *evaluated*. An organism's life is its *standard of value:* that which furthers its life is the *good,* and that which threatens it is the *evil.*[8]

Both the argument leading to the principle and the immediate subsequent context of the principle show that Rand "means what she says." The logical implication is clear: one's own life is of ultimate value to that person; all else, including the lives of others, is valuable only as a means. The principle is buttressed and further delineated by two others: "Happiness is that state of consciousness which proceeds from the achievement of one's values"[9]; and "To hold one's own life as one's ultimate value, and one's own happiness as one's highest purpose are two aspects of the same achievement."[10] Although Rand's essay does not have any explicitly stated principles of moral obligation, of what actions are morally right or wrong, one can deduce from the principles cited an ethical egoism equivalent to that of the strong version of the tough capitalist ethic extended to all aspects of life: one ought to do that which furthers one's own interests. There is no moral proscription from brutalizing someone else to enhance my own interests. Rand is logically committed to the ethic of selfishness for which she gained fame and notoriety.

Rand, however, can withstand the criticisms which shall later be directed against the tough capitalist ethic because she also maintains principles that jointly contradict the principles of her ethical egoism. What is valuable is not one's own life whatever its quality and whatever a person seeks to obtain as furthering his own life; only the right kind of life and the "right goals",[11] the values that are proper to man, "*qua* man", are valuable.[12] Rand, like Plato, Aristotle, Fromm, and many others solidly within the Western tradition of ethics, believes that because of our special human nature only a certain special kind of life is appropriate to a human being and only the furtherance of *that kind of life* is ethically permissible. The ultimate value for me is not my own life or happiness, but only my distinctively human life and happiness. Furthermore, since the ultimate value of a distinctively human life is what is common to all humans, *qua human,* " . . . every human living being is an end in himself . . ."[13] I am to enhance not only my own life as a human being, but the lives of others, *qua human* beings. Ethically, the solidarity of human beings is maintained and I am morally forbidden to treat other humans merely as means to my own life and happiness. By maintaining that only the true self, the self, *qua human,* is of ultimate value and that all selves, *qua human,* are for all humans ends in themselves, Rand contradicts her ethical principle of selfishness. Faced with criticisms of this principle, the defender of

Rand can take a stand on the non-egoistic principle that each person is to treat oneself and others equally as ends. On the other hand, if one argues that her ethic does not contain the virtue of selfishness, and is really quite traditional, a defender of the uniqueness of Rand's ethic can appeal to the principles of selfishness and maintain that Rand's ethic is really an egoism. To build a position around a logical contradiction is a nice way in the realm of ideas to have one's cake and eat it too, but at an intellectually great price.

Friedman's Obscure Ethic

Today Rand's writings have largely disappeared from popular discussion and the person most commonly identified with a *laissez faire* ideology is Milton Friedman, for over forty years the leader of the "Chicago School" of Economics, winner of the 1976 Nobel Prize for economics, now a senior fellow of the Hoover Institution, and together with his wife Rose, co-author of the recent best-seller, *Free to Choose*. This book, a series of television programs over the Public Broadcasting System, and a booklet, "The Economics of Freedom," grew out of a series of fifteen lectures delivered in 1977–1978.[14] These sources, the earlier book, *Capitalism and Freedom,* and Friedman's columns in *Newsweek* magazine are instances in which he writes or speaks as a philosopher advocating positions of *value* rather than as a relatively value-free scientist of economics. People who are familiar with any of these sources of Milton Friedman's "ethic," including his social–political philosophy, are almost certain to recognize him as a strong, persuasive advocate of a *laissez faire* ideology. In fact, the overall impression conveyed by *Capitalism and Freedom* and *Free to Choose* is that Hirshleifer's *laissez faire* ideology with its ethic of tough capitalism nicely summarizes Friedman's own position.

Here I shall only consider these two books, and before discussing to what extent Friedman accepts the tough capitalist ethic, I want briefly to comment upon that part of his *laissez faire* ideology that is now the center of public attention: his views as to the proper economic roles of government.

There are at least three reasons why those views deserve attention. The first is Friedman's approach. He does not try to establish some principles of the proper economic role of government and then use the principles automatically to approve policies that conform and to disapprove policies that violate the principles. Rather, he uses *laissez faire* principles of the proper economic role only as *guidelines* and tries, on a case-by-case basis, to draw up a balance sheet of the advantages of some actual or proposed government intervention. Whatever this approach may lose in being philosophically unsystematic, it gains in flexibility and helps to provide protection from doctrinaire thought. A second merit of Friedman's philosophical writing on government economic activity is that it sharply calls attention to the value of individuals not being told what to do by government. Over and over again he provides a challenge: is what is to be gained by some government economic activity sufficient to make up for the loss of freedom from government coercion? A third reason for studying Friedman's writings of social–political philosophy is that he plausibly argues that frequently the desirable effects sought by government intervention are not what one actually gets. For example, he cogently argues that the regulation of air, truck, and rail fares, instead of aiding the consumer by throttling monopoly power, has actually supported that power to the detriment of the consumer. Friedman's frequent challenge that government intervention does not obtain what its proponents want of it, his stress upon the value of being free from the government mandates, and his flexible approach to a *laissez faire* philosophy make his philosophical writings worth studying.

His writings, however, have to be read especially carefully and sceptically because they are frequently facile. By that I mean that what he writes about government's economic role is frequently vague or ambiguous, or it slides by crucial problems. Here I can only introduce examples without developing or documenting them.

Consider two examples from the recent *Free to Choose*. If the book is studied carefully, with special attention given to the passages listed under "Monopoly" in the index, one will see that Friedman's position for dealing with the extremely important problem of monopoly power is ambiguous. (Is he saying that the dropping of trade barriers would eliminate the need for antitrust laws?) Consider, secondly, Friedman's statement of principles of legitimate government economic intervention in the same book.[15] One of them includes the terms 'injustice' and 'justice,' but Friedman side-steps an elucidation of his concept of justice. Consequently, the sense of one of his own *laissez faire* principles is left vague. Important attacks upon *laissez faire* views are made because of claims about justice and because of allegedly inevitable monopoly power under conditions of mass production, and the least that Friedman needs to do is to make his own positions on these topics clear.

In *Capitalism and Freedom,* Friedman is very clear that he is opposed to any laws that prohibit racial discrimination in employment.

His reasons for his opposition, however, are not clear and they obscure his own *laissez faire* principles. He opposes antidiscrimination laws because: (1) racial discrimination, like musical preferences, is a matter of "taste"[16]; (2) the victims of racial discrimination suffer only "negative harm" and not "positive harm"[17]; and (3) if one grants government the right to prohibit racial discrimination, he must also grant government the right to require racial discrimination.[18] The principle of (3), however, seems to be absurd—except, perhaps, for a totalitarian or an anarchist. Does Friedman, in approving laws against fraud and murder (morally unjustified homicide), also grant government the right to require fraud or murder? Friedman ignores the problem. Although Friedman illustrates the difference between negative and positive harm and clearly implies that negative harm, unlike positive harm, cannot be used to justify government action, he does not make the distinction clear nor state clearly why he thinks that one, but not the other, is a justification for government action. Finally, if such an apparently important ethical matter as racial discrimination is a matter of taste, what other apparently important ethical matters are matters of taste? Are all matters of taste irrelevant for government intervention? Friedman ignores these problems. Friedman's use of principles (1)–(3) not only obscures his *laissez faire* principles of government intervention, it also illustrates the main philosophical flaw of both *Free to Choose* and *Capitalism and Freedom*.

To accomplish his main task in either book Friedman needs to utilize some value scheme of goods and evils. For his main task is to scrutinize government economic activity by making "a balance sheet, listing separately the advantages and disadvantages" of each case of actual or proposed intervention,[19] and to do that he needs some value scheme to determine what counts as either advantages or disadvantages and what weight is to be given to different goods or evils. But the careful reader of either book will be unable to state, for example, the relative worth that Friedman attaches to values such as freedom from government coercion, opportunity to earn a living, opportunity to develop and use one's human talents, and the freedom of economic security. (Friedman may not consider the latter to be a "freedom" or a good.) Whether intentional or not, Friedman fails to pay the price required for his kind of assessment of government economic policies and that is to be clear about the values he uses in making his assessments.

Let us now consider to what extent, if any, Friedman agrees with the tough capitalist ethic. That ethic, you may recall, is that in economic activities a person is morally required, or permitted, to pursue one's own interests without regard to the interests of others.

Because of the importance Friedman places upon the freedom of individuals to do what they want and, like Hirshleifer, upon the ability of a market economy to transfer the fulfillment of self-interests to the service of others, both books convey the general impression that he accepts the moral permissibility if not the obligatoriness of selfishness in economic matters. Nevertheless, Friedman, unlike Hirshleifer and Rand, does not *explicitly* try to summarize or develop an ethic and furthermore, there is no identifiable passage or set of passages in either book that rather clearly imply that unqualified selfish economic behavior is ethically permissible. Friedman almost always speaks of self-interested actions rather than of selfish actions, and, as we shall later see, to act in self-interest is not always to act selfishly. Thus, in placing importance upon the pursuit of self-interest, Friedman is not necessarily approving selfish economic behavior. In fact, in both *Free to Choose*[20] and *Capitalism and Freedom*[21] Friedman says he is using the term "self-interest" to mean whatever a person values whether one's values are materialistic or nonmaterialistic and whether they range from being altruistic to selfish—an unusually broad sense of the term. But if Friedman's emphasis upon the pursuit of self-interest does not logically imply his approval of acting without adequate regard for the interest of others, why does the general thrust of what he has to say about economic behavior suggest approval of acting this way?

The answer lies as much in what Friedman fails to say as in what he says. In both books, Friedman, like Hirshleifer, recognizes legal and market restraints upon the pursuit of self-interest, but he frequently fails to say anything about what Worthy calls "self-restraint." For example, when Friedman discusses what he means by self-interest in both books—see above for references—he is clear that he does *not identify* self-interest and selfishness, but he fails to add any proviso such as "of course, acting in a free market system provides no justification for acting selfishly." The lack of provisos that seem to be needed for clarification leave the impression that he agrees with Hirshleifer, "In this system everyone can be selfish. . . ."[22] Consider another example. The theme of the chapter on consumer protection in *Free to Choose* is that because of market restraints the pursuit of self-interests will generally better serve the consumer than will government regulations. He says that competition will protect the consumer "only because it is in the self-interest of the businessman to serve the consumer."[23] But Friedman nowhere in the chapter, even in his discussion of advertising and monopoly, says anything about self-imposed moral restraints in seeking self-interest—even when the avoidance of market restraints would enable a businessperson to make a killing at consumer expense. Other examples could be cited from either book. The construction is the same. What Friedman says does not clearly imply the proposition that in economic affairs a person is ethically permitted to act selfishly; his failure to add clarification or qualification to what he does say in both books suggests his agreement with the proposition. Consequently, readers of either book who are familiar with Hirshleifer's tough capitalist ethic are likely to believe that Friedman holds at least the weak version of that ethic.

There is, however, one statement in one of the books, *Capitalism and Freedom,* that clearly places a restriction upon the pursuit of selfish behavior. He says that the freedom of the individual to live "according to his own lights" is "subject only to the proviso that he not interfere with the freedom of other individuals. . . ."[24] The sentence from which I have just quoted is by far the clearest statement made in either book of Friedman's ethics of individual behavior. And in the clearest sentence made in either book in which he states what is good as an end Friedman identifies "freedom of the individual" as "the ultimate goal" for "judging social arrangements."[25] Surely, one would expect, Friedman defines and uses a clear concept of individual freedom. He does neither. In fact, examination of the two very important sentences from which I have just quoted together with their contexts reveals two different concepts of individual freedom that are incoherent with one another! Friedman's idea of individual freedom (or of voluntary transactions), which is central to both *Capitalism and Freedom* and *Free to Choose,* is too vague, too inchoate to make either book a plausible, cogent piece of Socio-Political-Economic Philosophy. And, more specifically to our immediate concern, Friedman's lack of a single definite concept of individual freedom renders his ethical restriction that one individual is not to interfere with another's freedom too opaque to know whether or not Friedman's philosophy is in agreement with even the weak version of Hirshleifer's tough capitalist ethic. Certainly the mere statement of an ethical restriction upon an individual's acting "according to his own lights" implies a limitation upon either version of the ethic of selfishness, but the nature of the limitation is only vaguely suggested.[26]

To suggest a proposition or a position is one thing; to clearly state or imply either is another. We earlier found that there are *suggestions* in both of Friedman's books of his agreement with at least the weak version of the tough capitalist ethic. We have just noticed a passage in *Capitalism and Freedom* that *suggests* disagreement with that version. Friedman does not have an ethical position about economic behavior that is sufficiently definite to logically evaluate, to logically accept, or cept, or to logically reject. He is no more clear about the moral right or wrong of economic behavior than he is about the human goods and evils by which he judges social arrangements. Rand's ethic avoids logical accountability by being built upon a contradiction. Friedman's ethic avoids it by being built upon obscurity. His is another way in the realm of ideas to have one's cake and eat it too. But the intellectual cost of his way may be as great as that of hers.[27]

The Allure of Tough Capitalism

Before we turn to a criticism of the tough capitalistic ethic, I want to say something about the attractiveness of the *laissez faire* ideology. I do this in the spirit of Hirshleifer's article, but because not all that I say can be inferred from it, I place my discussion in a separate section.

Let us assume that government's functions are restricted to maintaining free markets, to dealing with cost externalities, and to providing for collective choice goods. (Cost externalities are cases such as pollution, where a market cannot allocate to either the buyer or the seller the entire cost of a product or service. Collective choice goods are products or services where the market cannot allocate the products or services only to those willing and able to pay for them. National defense is a paradigm case of a collective choice good.) Thus, the only functions of government are to keep markets free and to distribute costs, products, or services that are not capable of being allocated by free markets. The beauty of such a limited government is that, first, people are free from government's deciding what is good or bad for them—people are free to pursue what they in fact actually value. Second, people are free from government coercion to determine how to obtain what they want. If people want oranges, rather than a government agency ordering land and labor into their production, the market provides an attraction for individuals, without their being ordered to do so, to set aside land and to labor to produce oranges. These two distinct kinds of freedoms claimed by the tough capitalism surely compel the hearts of many, however harshly some may judge the actual performance of a *laissez faire* system.

There are three other distinct great values that are promised by the ideology of *laissez faire*. Perhaps some readers are familiar with Read's story of a pencil, which Milton Friedman loves to cite.[28] No one can make a pencil. It is made of wood, of graphite, of tin, and of rubber. No one person knows how, starting from scratch, to produce all of its components. Literally thousands of people from different parts of the globe may have "cooperated" to attain the finished product from the raw materials. Free markets are to allocate the diverse thought, plant, and labor needed to produce the material, the final product, and its delivery to a person who happens to want a pencil. Otherwise thousands of hours of work would be required to decide how much tin, for example, is going to be produced, how much of the tin produced is going to be used for pencils, who is going to work the tin mines, who is going to work at what job at a refinery, and so on. Secondly, this marvelous allocating function of the markets can be accomplished to yield a great cooperative-like activity with each person of the activity following his/her own interests, but without the person's knowing or caring about others. As Hirshleifer suggests, each person can be as selfish as he/she pleases to produce unintentionally what I and others may happen to want. Thirdly, with each person acting uninhibitedly selfishly, great energies are released to maximize unintentionally what others want.

Freedom from government's dictating to people what they ought to want, manifold freedom from government's ordering people about to obtain what people actually want, and the tripartite efficiencies of the allocation function of free markets constitute the utopian-like promises of the most commonly held concept of a *laissez faire* system. These values have a wide appeal, and no wonder that people with some business management or economics can read Hirshleifer's "Capitalist Ethics—Tough or Soft" and respond to the articles as though it were a manifesto for a pluralistic utopia within the limits of imperfect human nature. We philosophers who want a mutually enriching dialog with businesspeople need to recognize and appreciate these values attributed to a *laissez faire* system. For they affect attitudes toward a wide variety of more specific issues in Business Ethics.

The Futility of the Laissez Faire Ethic

In this section let us evaluate the ethic of tough capitalism and the claim that because free markets divert the effects of selfish economic actions to the well-being of others there is a special, less stringent ethic of economic activities.

Our evaluation must bear in mind an important distinction. Self-interested actions and selfish actions are not the same. Selfish actions are always self-interested, but not vice versa. The former are self-interested and also fail to adequately heed some interest of another person. Suppose that I stop by a candy counter to get a candy bar my wife has been anticipating and expecting me to bring home; I get an urge for some gum for myself and spend the last quarter in my pocket for the gum. My action is selfish. On the other hand, suppose that I buy some gum simply because I want some, but in doing so I do not neglect any interest due to another. My action is self-interested without being selfish. In evaluating an ethic that permits selfishness we are evaluating an ethic that permits more than unselfish, self-interested actions.

Is the tough capitalist ethic an adequate ethic for economic affairs? Here we can best answer the question by considering cases. Suppose that a young man who knows that in two months he is to begin service in the Army seeks a job requiring considerable training and he knows that he would not obtain the job were he to tell his employer of his temporary employment situation. We can easily imagine circumstances, such as the person's seeking a job in an anonymous metropolitan area, in which there would be no effective market restraints to hinder his selfishly taking the job even though he knew that his doing so would be a considerable loss to the employer. So long as the young person can bypass legal and market restraints there is, according to the ethic of tough capitalism, no moral restraint to inhibit the young man from pursuing his selfish interests at the expense of the interests of an employer. (The tough capitalist ethic does not obligate a person to put oneself under either legal or market restraints. On the contrary, in economic matters one is obligated (or permitted) to avoid either kind of restraint if doing so will maximize self-interest.) Imagine another case. A product manufacturer finds that he or she can maximize profits by combining the production of a poor product with skillful advertising that skirts legal restraints while it creates a quasimonopolistic demand for the product. Although the manufacturer may know that self-interests are being gained at the expenses of others, there is no moral prohibition in the ethic of *laissez faire* capitalism from doing so. Or consider a case of a recent widow taking herself and her young children from Kalamazoo to her parents' home in San Francisco. Her battered car breaks down in Chicago and she needs another. There is no moral restraint to hinder a salesperson from selling her a car that the dealer knows to have a bad oil pump and to be almost certain to break down before the mother and her children can cross Iowa. In fact, there is nothing in the tough capitalist ethic according to which the salesperson would do anything wrong knowingly to sell the woman a car with highly dangerous brakes. The only relevant "moral" questions in producing or selling dangerous products are: is it profitable for me, and can I get away with it? Market restraints and legal restraints allowable under *laissez faire* capitalism simply do not prevent a person's obtaining small economic gain at the expense of considerable loss and suffering to others. In fact, if my reader will only take time to consider the implications of the *ethic* of tough capitalism, he or she will see that it *morally* requires (or permits) killing another person for one's own small economic gain. And, unless my reader is committed to the position that in economic matters might makes moral right, he or she will agree that Hirshleifer's *laissez faire* ethic is far removed from being an adequate ethic for business or any other economic relations.

A *laissez faire* advocate may try to salvage the tough capitalist ethic by arguing that, although there would be some cases under a *laissez faire* system in which the pursuit of selfishness leads to great harm of others, that system with its ethic is superior to other systems or ethics because free markets under tough capitalism for the *most part divert* selfish pursuits to the service of the interests of others.

My rebuttal to this reply must be brief, and I can only sketch four points that undermine it. First, consider the lack of market restraints upon self-interest because of the existence of monopolies and oligopolies. The former seem to be essential for efficient utilities and modes of transportation. Oligopolies, the dominance of a market by a few, seem to be essential for the efficiencies of mass production. (Almost all products today are produced under the condition of oligopoly rather than under free competition.) The devastating labor monopoly power of organized unions can hardly be avoided given the political realities of most Western countries. Even rather small businesses frequently constitute oligopolies. Four retailers of lumber serving an isolated population area of 70,000 have the power of oligopoly. Perhaps the defender of the claim that free markets will generally divert selfishness to the service of others can deal with the many stubborn facts of the lack of free markets, but the defense needs to recognize that the invisible hand of Adam Smith seems to be arthritic and that the burden of argument rests with it.

Second, the avoidance of the restraints of markets without having established a monopoly or oligopoly is not an extremely rare fact of economic life. Sometimes the conditions of mobility essential to free markets enable buyers or sellers to avoid market restraints. Only a few hours before composing this page I heard a radio report by a government official who warned people not to deal with some unscrupulous door-to-door salespeople who work for a company that yearly changes its name and the locations of sales personnel. The defender of the claim that markets are a sufficient condition generally for allocating selfish pursuits to service of others needs to be realistic in assessing how much harm done by avoiding market restraints he or she is willing to tolerate. And this brings me to my next point.

Third, in looking for examples of a *laissez faire* system diverting selfishness to general well-being, it is easy to overlook the effect of the actual legal restraints that are contrary to the well-known *laissez faire* principle that government economic activity is to be confined to preserving price competition and to providing for collective choice goods and for cost externalities. The local government official who warned people of the apparently deceitful door-to-door salespeople was violating that principle—unless the concept of preserving price competition is unmanageably broad. Laws against selling heroin even to children certainly violate the principle. So do laws that enable consumers to sue sellers of goods or services for negligence. The threat of government coercion as well as the market actually "forces" sellers to provide adequate goods or services. Furthermore, it is doubtful whether laws considered essential to basic law and order can be justified on the well-known *laissez faire* principle. Protection from murder could be marketed; it is not a collective choice good. Government probably has to make some value decisions of what is good or bad, or what is right or wrong, if it is to be responsible for the fundamentals of law and order. (This is one of the reasons, I suspect, why Friedman holds any *laissez faire* principle only as a guideline and admits that in some cases government is to do for people what they cannot do individually so well for themselves.) In any case, were it not for actual government restraints that are contrary to the most commonly recognized *laissez faire* principle, the markets would not be thought to be sufficient generally to divert selfish economic pursuits to the service of others.

Finally, the plausibility of the claim that markets are adequate to produce good for others from selfish actions depends upon the fact that people generally are not as selfish as Hirshleifer claims them to be. If people were as selfish as Hirshleifer and others imply, most of us would be as unscrupulous as the people of organized crime, who are the exemplars of the strong version of the tough capitalistic ethic. *To some extent,* of course, markets do allocate to the service of others the effects of actions that are to *some extent* selfish. However, without some of the self-restraint on the part of most people, the markets would be seen to be quite inadequate generally to allocate the effects of completely selfish actions to the service of others.

For the four reasons just stated, I deny that free markets are sufficient generally to convert selfishness to the service of others, and accordingly, I deny that this claim justifies the position that there is a special, less stringent business ethic than an ethic applicable to merely personal relations. And this position, of course, cannot be justified by appeal to the claim that people are always selfish, for if that justified a selfish business ethic, it would also justify an ethic of selfishness for nonbusiness activities. There would be no special, less stringent business ethic. The most commonly recognized and the most frequently accepted version of the *laissez faire* ideology does not provide an adequate rationale for the position that there is a special business ethic.

Concluding Remarks

The most frequent objection that I have met against my criticism of Hirshleifer is that I distort his view because I fail to notice that he is assuming that commonly recognized standards of decency will be applied in economic activities. I am charged with distorting the ethic of tough capitalism by claiming that it justifies homicide, fraud, or even lack of candor in allowing a person to sell the auto with the bad oil pump to our widow who needs to get her family on their way to San Francisco. The ethic of tough capitalism, my critic says, assumes that all actions will be circumscribed by commonly accepted canons of morality, and does not morally permit the obviously immoral actions that I have attributed to it.

This objection to my criticism of the ethic of tough capitalism will not do. In the first place, the objection is not warranted by the text of Hirshleifer's article. But no matter. I suspect that he and many who want to defend his tough capitalist ethic do assume that it, with its implications, is bounded by commonly accepted standards of decency. However, if this is the case, the ethic of tough capitalism differs only in name from the ethic of soft capitalism. For the latter only requires that in economic activities, as in non-economic activities, one give due weight to the interest of others. After all, that ethic does not require that business persons act altruistically; business activities are only subject to the self-restraint that one is not to act selfishly. If the tough capitalist ethic is made morally plausible by adding to it the morality of common decency, it would be indistinguishable from the soft ethic.

This does not mean that I have shown that there is no adequate ethic distinctive of economic matters and less stringent than that of the soft capitalist ethic. My arguments do not show and are not intended to show the *impossibility* of that. I have attacked what I think is the most commonly recognized special business ethic and the most familiar way of justifying that there is such an ethic. I want to close by making three comments upon the possibility of developing a distinctively less stringent business ethic.

First, such an ethic cannot be clearly and cogently developed as an ethic of selfishness with some provisos tacked onto it. Unless the term "selfish" is used in a confusing way, to act selfishly is to act in some degree immorally. Standards for selfishness may be more lax in economic than in non-economic affairs, but that does not morally justify any economically selfish actions. Second, the stringency of moral responsibilities varies with the character and the degree of personal relationships. (Normally the loneliness of one's single parent imposes a responsibility not imposed by the loneliness of others one may happen to know. Sometimes personal relations in economic affairs do the same.) Nevertheless, most of our relations to others in economic affairs are highly impersonal and this is probably why some people think that there is a less stringent ethic for economic affairs. I suspect that a formidable, competent effort to develop a special business ethic will conclude that there is only one set of ethical principles common to all human affairs, but with a difference of moral stringencies being applicable to a difference of personal relationships.

Third, consider a paradox. My reader has probably heard that the greater political freedom allowed individual citizens, the greater the political responsibility required of individual citizens. In general, the common saying is that the greater one's freedom, the greater her or his responsibility. Paradoxically, the call for a special ethic of minimal moral responsibility in business has come from those most strongly advocating a maximum of freedom in economic affairs. (That is the underlying reason why the Hirshleifers, the Rands, and the Friedmans are accused—quite unfairly, I believe—of being spokespersons for the economically powerful.) However, I suggest that my reader who tries to construct an ethic of business will find that the greater the business freedom an individual is assumed to have, the greater the moral responsibility applicable to the individual as a business person. "Capitalist Ethics—Tough or Soft" deserves judiciously critical thought that does not confuse freedom from government coercion in business with freedom from moral responsibility in business. The greater that freedom, the greater that responsibility.

Notes and References

[1]Adam Smith, *An Inquiry into the Nature and Causes of the Wealth of Nations*, R. H. Campbell, A. S. Skinner, and W. B. Todd, eds., in two volumes (Oxford: Oxford University Press, 1976), pp. 26–27. The first volume of the original edition was published in London in 1776.

[2]Jack Hirshleifer, "Capitalist Ethics—Tough or Soft?" *The Journal of Law and Economics* 2 (1959), 114–119.

[3]James C. Worthy, "Religion and Its Role in the World of Business," *Journal of Business* 31 (1958): 293–303. Hirschleifer very accurately summarizes the portion of this paper that is relevant to his own article.

[4]Hirshleifer, *op. cit.*, p. 117.

[5]Hirshleifer briefly indicates why both tough and soft socialism must be rejected.

[6]*Ibid.*, p. 118.

[7]Ayn Rand, *The Virtue of Selfishness: A New Concept of Egoism,* (New York: The New American Library, 1964).

[8]*Ibid.*, p. 17.

[9]*Ibid.*, p. 28.

[10]*Ibid.*, p. 29.

[11]*Ibid.*, p. 22.

[12]*Ibid.*, p. 23.

[13]*Ibid.*, p. 27.

[14]Milton and Rose D. Friedman, *Free to Choose: A Personal Statement* (New York: Harcourt, Brace, Jovanich, 1980). The fifteen lectures, with a question period following each, were given at various locations in 1977–1978. Edited versions of the lectures and question periods are available on videotape from Harcourt, Brace, Jovanovich, New York: 1980. The booklet, ''The Economics of Freedom, '' is published by the Standard Oil Company of Ohio, Cleveland: 1978. *Capitalism and Freedom* (Chicago: University of Chicago Press, 1962) is Milton Friedman's predecessor work to *Free to Choose,* and was written with the assistance of his spouse, Rose. Because the positions of Milton are of interest in this paper, I shall refer to ''Friedman's'' views, meaning those of Milton, rather than to the views of the Friedmans.

[15]Friedman, *Free to Choose, op. cit.,* pp. 28–33.

[16]Friedman, *Capitalism and Freedom,* op. cit., pp. 110–113.

[17]*Ibid.*, p. 112–113.

[18]*Ibid.*, p. 113.

[19]*Ibid.*, p. 32; quoted by Friedman, *Free to Choose, op. cit.*, p. ix.

[20]*Op.cit.*, p. 27.

[21]*Op.cit.*, p. 200.

[22]Hirshleifer, *op. cit.*, p. 118.

[23]Friedman, *Free to Choose,* op. cit., p. 222.

[24]Friedman, *Capitalism and Freedom,* op. cit., p. 195.

[25]*Ibid.*, p. 12.

[26]A minimum that Friedman means by ''individual freedom'' (or ''voluntary transaction'') is that an individual is free, or a transaction voluntary, only if the person is unrestrained by physical force—whether by its use or the threat of its use. Consequently, it seems that there is a clear minimal restriction that Friedman places upon both selfish and non-selfish behavior: an individual is not ethically permitted to restrain another by physical force. Friedman, however, does not unambiguously maintain this restriction. For he clearly grants that restraint by physical force is sometimes justified, but he does not provide any definite principles of its justification.

[27]In rebuttal of my charge, a defender of Friedman could reply that Friedman's ethic seems to be vague, ambiguous, or in some other way facile by arguing that Friedman should not be expected to have clear positions about human goods or morally right actions because Friedman maintains some kind of ethical relativism. If a person is sceptical about ethics, why should that person be held responsible for having a clear ethic? This rebuttal of my charge, however, will not do. I shall mention only one reason. That Friedman is some kind of ethical relativist can be supported by citing two sentences from page 12 of *Capitalism and Freedom* and pages 106 and 135 of *Free to Choose*. But Friedman does not clearly imply any ethical relativism in either book. There is only the suggestion, the flirtation with, not the statement of clear implication of any kind of relativism. Friedman's ethical relativism cannot provide an adequate defense of the obscurity of his ethics because that too is obscure.

[28]Friedman, *Free to Choose, op. cit.,* pp. 11–12.

The Business of Ethics

Norman Chase Gillespie

It is the business of ethics to tell us what are our duties, or by what test we may know them.

John Stuart Mill
Utilitarianism

The public image of business does not always inspire public confidence, since it is often assumed that talk of ethics in business is only talk, not something that makes a difference in practice. Business executives are pragmatic individuals, accustomed to dealing with their environment as they find it and not inclined to question how things ought to be. That frame of mind reinforces the public image of business as impervious to moral imperatives.

That image is often only confirmed in the press, for instance, by such articles as those of Albert Carr, which embrace the purest kind of moral conventionalism: that which is generally done in business sets the standard of ethical conduct, so that an executive acts ethically as long as he or she conforms to the general practice. Carr goes so far as to maintain that misrepresentation in business is as ethical as bluffing in poker, and that only needless concern and anxiety will result from applying the ordinary moral standards of society to the conduct of business.[1] On this score, I believe Carr is completely mistaken.

This paper will argue that ordinary moral standards do apply to business decisions and practices and will explain *how* they apply. This should result in a clearer picture of the relationship between business and ethics—what it is now and what it ought to be.

Carr, in setting forth the conventionalist position, argues:

(1) Business, like poker, is a form of competition.

(2) In this competition, the rules are different from what they are in ordinary social dealings.

(3) Anyone who abides by ordinary moral standards instead of the rules of business is placed at a decided disadvantage. Therefore,

(4) It is not unethical or immoral to abide by the current rules of business. (These rules are determined in part by what is generally done in business and in part by legal statutes governing business activities.)

In support of this position, three reasons might be offered:

(1) If a business practice is not illegal, it is thereby ethically acceptable.

(2) If a businessman does not take advantage of a legal opportunity, others will surely do so.

(3) If a practice is so widespread as to constitute the norm, everyone expects conformity.

The claim that it is ethically correct to do something because it is not illegal is, of course, one of the conventionalist's weakest arguments, since it should be obvious that legality does not establish morality—it may not be illegal for a teacher to favor some students over others for non-academic reasons, yet it is clearly unethical. When one speaks of ethics in business, it is to establish what business practices ought to be. The law, as written, does not settle that issue. The other two reasons, however, may appear to have some merit and require more detailed analysis.

Business as a Game

Suppose that such things as industrial espionage, deception of customers, and shading the truth in published financial statements are common enough to be of broad concern, in effect comprising some *de facto* state of business affairs. What bearing would such a state have upon what is moral or ethical in conducting business? Would the existence of such "rules of the game" relieve owners, managers, and employees of otherwise appropriate ethical obligations? Or, would such behavior merely be a matter of business strategy and not a matter of ethics?

The obvious fallacy in the "business-as-a-game" idea is that, unlike poker, business is not a game. People's lives, their well-being, their plans, and their futures often depend upon business and the way it is conducted. Indeed, people usually exchange part of their lives (i.e., the portion spent earning money) for certain goods and services. They have the right not to be misled or deceived about the true nature of those goods or services. Similarly, elected officials have a duty to legislate and act for the good of their country (or state). It can hardly be right for business executives to frustrate them in the performance of that duty by providing them with evasive answers or by concealing relevant facts.

The Price of Duty

So, the poker analogy, while informative of the way things *are,* seems to have no bearing at all on the way they *ought* to be in business. Why, then, do so many people adopt the conventionalist position that "business is business and, when in business do as the others do"? Some take that position for essentially the same reason Yossarian offers to justify his conduct in the novel, *Catch 22*. Yossarian has refused to fly any more combat missions and when asked, "But suppose everybody on our side felt that way?" he replies, "Then I'd be a damn fool to feel any other way. Wouldn't I?"[2] If everyone were refusing to fly, Yossarian says, he would be a fool to fly. In business, the position would be: if everyone is bluffing, an individual would be a fool not to do the same. On this point, Yossarian and the conventionalist are correct, but not because there are special rules (or special ethics) for airplane gunners and people in business. The reason, instead, is that our ordinary moral reasoning does, indeed, make allowance for just such cases. In other words, the idea that there is something ethically distinctive about a situation in which a person in business may find him- or herself is sound. But it is sound because ordinary moral reasoning allows for such circumstances, not because there are special ethical rules for people in business comparable to the rules of poker.

The sort of considerations I have in mind all involve the *cost* of doing what would normally be one's duty. There are at least three ways in which a normal or ordinary duty may cease to be so because the cost is too high. The first of these is widely recognized: sometimes the *moral cost* of obeying a standard moral rule is too great, so one must make an exception to that rule. If the only way to save someone's life is by telling a lie, then one should normally lie. If treating an accident victim involves breaking a promise to meet someone on time, then one should normally be late. In a variety of circumstances obeying a moral rule might require breaking some other, more urgent, moral duty. In these circumstances, the more urgent duty dictates an exception to the lesser rule.

The second way in which an ordinary duty may cease to be a duty is when the *cost to the individual* of fulfilling that duty is too high. For example, when driving an automobile, one normally has the duty not to run into other cars, and one also has the general duty not to harm or injure other persons. But suppose one is driving down a steep mountain road and the brakes fail. One might have a choice among three options: cross into the oncoming lane of traffic, go off the cliff on one's right, or drive into the car in front. In such a case, a driver would not act wrongly by choosing the third option, even though there is a way in which the duty of not injuring others and not driving into other people's cars can be met, namely, by going off the cliff. In these circumstances, the cost to the individual of meeting the duty is simply too high, and virtually no one would blame the driver or condemn the action as morally wrong if he or she drove into the back of the car ahead rather than going off the cliff.

The third way in which a normal duty may turn out not to be a duty is the kind of situation described by Yossarian. If everyone else is not doing what ought to be done, then one would be a fool to act differently. This third consideration does not obviate all duties, e.g., just because everyone else is committing murder does not make it right for you to do so, but it does apply to those cases in which the *morally desirable state of affairs can be produced only by everyone, or virtually everyone, doing his or her part*. With respect to such a duty, e.g., jury duty, one person alone cannot accomplish anything; one can only be placed at a disadvantage vis-à-vis everyone else by doing what everyone ought to do but is not doing. This sort of situation can be described as a "state of nature situation,"[3] and by that I mean a situation in which certain moral rules are generally disobeyed either by everyone or by the members of a well-defined group.

In dealing with such situations, the fact that other people can be expected to act in certain ways is morally relevant in that it creates a special sort of moral dilemma. If one does what everyone ought to do but is not doing, then one will, in all likelihood, be at a disadvantage. The morally questionable behavior of others creates the circumstances in which one finds oneself and in those circumstances it may be necessary to fight fire with fire and to resist deception with deception. But replying in kind only prolongs the state of nature situation, so one's primary goal should be to attempt to change the situation. No one ought to take unfair advantage of others, but no one is obligated to let others take unfair advantage of him or her.

It is absolutely essential to note, in connection with such situations, that people are not doing what they ought to be doing. The conventionalist recognizes that simple application of ordinary moral rules to such situations is inadequate. But it is a mistake to conclude (1) that ordinary rules do not apply *at all* to such cases, and (2) that business has its own distinctive set of rules that determines one's duties in such circumstances. Both points are incorrect because (1) the ordinary rules help define the situation as one in which people are not doing what they ought to be doing (we apply the ordinary moral rules to such cases and find that they are not being generally observed), and (2) the considerations that are relevant in determining one's duties in such circumstances do not constitute a special set of factors that are relevant only in business. The mitigating considerations apply generally and are an important part of ordinary moral reasoning.

When virtually everyone is not doing what ought to be done, it affects what we can morally expect of any one individual. That person does not have a duty to "buck the tide" if doing so will cause the individual substantial harm or not do any good. But, in conjunction with everyone else, the person *ought to be* acting differently. So the tension one may feel between what one does and what one ought to do is quite real and entirely appropriate.

In the conventionalist argument, these two considerations—the cost to the individual, and what everyone else is doing—recur again and again: It is right to lie about one's age and one's magazine preferences when doing otherwise will prevent you from getting a job; right to engage in industrial espionage because everyone else is doing it; right to sell a popular mouthwash with a possibly deleterious form of alcohol in it because cigarettes are sold to the public; and right to sell master automobile keys through the mails (to potential criminals) because guns are sold.[4]

Of these four examples, the last two seem to me to be clearly wrong since neither the high cost to the individual of doing otherwise nor the existence of a general practice has been established. Industrial espionage, however, is a good illustration of a "state of nature situation," and if (1) one does it to others who are doing the same, and (2) it is necessary to "fight fire with fire" for the sake of survival, then it would not be morally wrong. For the job applicant, the conditions themselves are morally dubious, so here, too, it may be a case of fighting fire with fire for the sake of personal survival. But notice, in each of these examples, how distasteful the action in question is; most of us would prefer not to engage in such activities. The point is that conditions may be such that the cost of not engaging in them may be so great that an individual caught in such circumstances is blameless. At the same time, however, we do feel that *the circumstances* should be different.

The second consideration, distinct from the cost to the individual, is that one person doing what everyone ought to do (but is not doing) will accomplish nothing. This can be the case even where the individual cost is insignificant. To take a homely example, suppose there were a well-defined path across the local courthouse lawn as the result of shortcuts taken across it. It would not cost anyone very much to walk around instead of across the lawn, but if one knows his fellow citizens and knows that the path is there to stay, then walking around will accomplish nothing. So one may as well take the path unless, of course, one decides to set an example of how others ought to be acting. Since it costs so little, it might well be a good idea to set such an example. This would be one small way at least of trying to change the situation.

Although one's primary goal ought to be to change the situation, that statement like all claims about what one ought to do, is subject to the moral precept that individuals have a duty to do only what they can do. So, if it is impossible for one individual to change the situation, that person does not have a duty to change it. What is true is that the situation ought to be different, but to make it so may require the combined efforts of many people. All of them collectively have the duty to change it, so this is not a duty that falls solely or directly on the shoulders of any one person. For the individual executive, then, the question is primarily one of what he or she in conjunction with others can accomplish. Secondarily, it is a question of the likely personal cost to the executive of instituting or proposing needed changes.

A Role for the Individual Executive

At the very least, executives should not *thwart* the impetus for change on the ground that business sets its own ethical standards. Everyone has a legitimate interest in the way business is run, and Better Business Bureaus and legislative inquiries should be viewed as important instruments serving that interest. We know on the basis of ordinary moral rules that in certain business environments a new way of acting is a desirable goal. If no one else will join in the promotion of that goal, then the individual executive can, as the poet said, "only stand and wait." But according to that same poet, John Milton, "They also serve who only stand and wait."

The essential difference between the conventionalist position and ordinary moral reasoning comes out most clearly in the following example, provided by Carr in defense of his position. A businessman, Tom, is asked by an important customer in the middle of a sales talk to contribute to the election campaign of a candidate Tom does not support. He does so, and the talk continues with enthusiasm. Later, Tom mentions his action to his wife, Mary, and she is furious. They discuss the situation and the conversation concludes with her saying, "Tom, something is wrong with business when a man is forced to choose between his family's security and his moral obligation . . . It's easy for me to say you should have stood up to him—but if you had, you might have felt you were betraying me and the kids. I'm sorry you did it Tom, but I can't blame you. Something is wrong with business."

Carr comments that, "This wife saw the problem in terms of moral obligation as conceived in private life; her husband saw it as a matter of game strategy." Those who would refuse to make the contribution "merit our respect—but as private individuals, not as businessmen."[5]

What Tom did was not morally wrong in those circumstances, but not for the reasons cited in Carr's paper. There is something wrong with *the situation* in which Tom found himself. It *ought not be the case* that one has to choose between one's family and being honest about one's political preferences. Carr fails to recognize this, and either misses or ignores entirely the fact that Mary makes precisely this point: she does not blame her husband or say that he did the wrong thing in those circumstances; what she says, instead, is that *something is wrong with business* when a person has to act as her husband did. It is business and the way it is conducted that ought to be changed. The conventionalist position simply blocks out such an issue: it nowhere considers how business ought to be. It merely says that "the way it is" is all that need be taken into account in deciding what would be ethical.

An analogous situation exists in connection with the financing of political campaigns. No one blames candidates for taking contributions from lobbyists and other individuals since they need the money to run for office. But many people do think that the system ought to be changed. In other words, the current practices are not as honest or ethical as they ought to be. Now, how can the conventionalist handle such a claim? It seems obvious that he cannot, since he systematically rules out applying ordinary moral standards to business practices. But the correct position is that these standards do apply, and sometimes we find they are not being put into practice. In precisely those cases, the general practice ought to change.

The Need for Change

There is a most important difference, then, between asking "What are the individual duties of a person doing business?" and "What are the ways in which business ought to be conducted?" Both are an essential part of the ethics of business but the conventionalist simply ignores the second question in attempting to answer the first. The answers to the second question can be found, for the most part, by consulting our ordinary moral standards of how people ought to act vis-à-vis one another. When we find that business is not as moral as we would like it to be, that *does* have some bearing upon the answer to the first question. But, as I have argued in this paper, ordinary moral reasoning is prepared to take those facts into account. It is not at all necessary to postulate a special ethical outlook or a distinctive set of ethical rules for business in order to explain the ethical relevance of such phenomena to the individual businessman.

Ordinary moral reasoning, then, is far richer than mere conventionalism and the factors it takes into account are relevant in many managerial and executive decisions. Ethics can be subtle, as well as realistic, and conventionalism is unrealistic when it obscures the moral imperative for change.

NOTES AND REFERENCES

[1]Albert Carr advocates conventionalism in *Business as a Game* (New York, 1968); "Is Business Bluffing Ethical?" *Harvard Business Review* (January-February, 1968) pp. 143–153; and reiterates it in "Can an Executive Afford a Conscience?", *Harvard Business Review* (July-August, 1970), pp. 58–64. In defending himself against the criticism that he is condoning unethical behavior, Carr insists that an executive who acts according to prevailing business practices "is guilty of nothing more than conformity; he is merely playing the game according to the rules." "Showdown on Business Bluffing," *Harvard Business Review*, (May-June, 1968). p. 169.

[2]Joseph Heller, *Catch 22*, (New York, Dell, 1961), ch. ix.

[3]Marcus G. Singer uses this term in *Generalization in Ethics* (New York: Knopf, 1961), pp. 153 and 156–157.

[4]Examples from Carr's "Is Business Bluffing Ethical?", pp. 144, 146, and 148.

[5]*Ibid.*, pp. 152–153.

The Moral Status of Bluffing and Deception in Business

Thomas L. Carson and Richard E. Wokutch

In recent years, the legality and morality of many business practices have come under increasing scrutiny. A range of activities involving exaggeration and deception in advertising, selling, and negotiating has been particularly controversial. This controversy was intensified by a recent Wall Street Journal article[1] concerning a negotiations course taught by Professor Howard Raiffa at Harvard Business School. In this course, students were graded on the settlements they negotiated with other students in various contrived business situations. It was possible for the students to misrepresent factual information and their negotiating positions to improve their settlements and in turn, their grades.

In the first section of the paper, we will consider some examples of what is commonly referred to as bluffing in business negotiations and attempt to determine whether or not they involve lying. We will argue that bluffing typically does involve lying. The second part of the paper considers some of the various economic incentives and disincentives for lying. We will argue that bluffing and other deceptive practices are often profitable and sometimes felt to be economically necessary. Then we will attempt to determine whether this fact provides any kind of *moral* justification for such practices. We will also attempt to determine whether or not the fact that such practices are generally condoned makes them morally acceptable.

Is Bluffing Lying?

Suppose that I am trying to sell my home. I need to sell it soon and I am willing to take as little as $45,000 for it, if that is the best price I can get. A potential buyer comes to me and I ask for $55,000. The buyer refuses this and offers me $45,000. At this point I say that my absolute lowest selling price is $50,000. The buyer then takes it for $50,000. Have I told a lie? Consider the following three variations on this case. Case #2: I make my bluff and tell the buyer that $50,000 is my lowest selling price. The buyer then offers me $47,000 and I agree to that price. Case #3: This is just like cases #1 and #2 except that after the offer of $45,000 I falsely claim that someone else has offered me $50,000 for the house, and I say that I will let it go for $50,000 as a personal favor. The buyer then rapidly takes the house for $50,000. In case #4, instead of simply making up a story about another prospective buyer, I have my cousin come in and pretend to make me an offer of $50,000 for the house. Cases #3 and #4 clearly involve lying, on my part in #3, and on the part of my cousin in #4. Almost everyone would agree that my behavior in these two cases is immoral and possibly illegal. However, the sort of bluffing involved in cases #1 and #2 is very common in business and thought by many to be morally acceptable.

In the context of labor negotiations it is also difficult to distinguish between bluffing that involves lying and bluffing that does not. Most would agree that the following constitute lies:

(1) Management negotiators misstating the profitability of a subsidiary to convince the union negotiationg with it that the subsidiary would go out of business if management acceded to union wage demands.
(2) Union officials misreporting the size of the union strike fund to portray a greater ability to strike than is actually the case.

The following cases, however, are not quite so clear as to whether or not they constitute lying:

(1) Management negotiators saying, "We can't afford this agreement," when it would not put the firm out of business, but only reduce profits from somewhat above to somewhat below the industry average.
(2) Union negotiatiors saying, "The union membership is adamant on this issue," when they know that while one half of the membership is adamant, the other half couldn't care less.
(3) Union negotiators saying, "If you include this provision, we'll get membership approval of the contract," when they know they'll have an uphill battle for approval even with the provision.

What is lying? A lie must be a false statement. Arnold Isenberg, however, holds that a statement can be a lie if one does not believe it.[2] This is most implausible. For if what one says is true, that is always sufficient to defeat the claim that one is lying. However, we can say of a person who makes true statements that are believed to be false that he or she *intends to lie,* which is perhaps just as blameworthy as lying.

A lie must be a false statement, but not all false statements are lies. If I am a salesman and say that my product is the best on the market *and sincerely believe this to be the case,* my statement is not a lie, even if it is untrue. A false statement is not a lie unless it is somehow deliberate or intentional. But what exactly does this mean? Must one believe that what one says is false, or is it enough that one not believe it? Chisholm and Feehan hold that the latter is all that is necessary.[3] This makes the concept of lying somewhat broader than it would otherwise be. Suppose that I tell you that mine is the only store in town that sells a certain product with the intent that you believe it. Suppose also that what I say is false. What additional conditions must be satisfied in order for us to say that I am lying? Many would hold that in order for my statement to constitute a lie I must believe that it is false. Chisholm and Feehan would say that all that is required is that I not believe what I say. Our own intuitions about these cases are not clear enough to favor one view over the other. Whether or not making a false statement that one does not believe to be true, but that one does not believe to be false either constitutes a genuine case of lying, it is not significantly less morally objectionable than making a false statement that one believes to be false.

Suppose that we define a lie as an intentional false statement and let the expression "intentional false statement" be ambiguous between "false statement that the speaker (writer, etc.) believes to be false" and "false statement that the speaker does not believe." According to this definition, I am telling a lie when I say, "This aftershave will make you feel like a million bucks." This definition implies that we lie when we exaggerate, e.g., a negotiator representing union workers making $10/hour but seeking a substantial raise says, "These are slave wages you're paying us." When I greatly exaggerate or say something in jest, I know that it is very improbable that the other person(s) will believe what I say. The reason that our two present examples do not seem to be cases of lying is that they do not involve the intent to deceive. This suggests the following definition of lying:

(1) A lie is a deliberate false statement intended to deceive another person.

This definition is inadequate in cases in which a person is compelled to make false statements. For example, I may lie as a witness to a jury for fear of being killed by the accused. But it does not follow that I hope or intend to deceive them.[4] I may hope that my statements do not deceive anyone. We might say that what makes my statements lies is that I realize or foresee that they are likely to deceive others. This suggests the following definition of lying:

> (2) A lie is a deliberate false statement that is thought to be likely to deceive others by the person who makes it.

This definition is also lacking because a person can lie even if he or she has almost no hope of being believed. A criminal protesting his or her innocence in court is lying no matter how unlikely it is that the criminal thinks anyone else will be convinced. The following definition is more plausible than either (1) or (2):

> (3) A lie is a deliberate false statement that is either intended to deceive others or foreseen to be likely to deceive others.

It appears that this definition implies that our first two examples of business bluffing constitute lies. In each case, I am making a deliberate false statement with the intent to deceive the other party into thinking that my minimum selling price is greater than $45,000. Here one might reply that this need not be my intent. No one familiar with standard negotiating practices is likely to take at face value statements that a person makes about his or her bottom price. One might argue that in the two cases in question, I intend and expect my statement that $50,000 is my lowest possible price to be taken to mean my lowest possible price is something around $45,000. If this is my intention and expectation, then my bluffing does not constitute a lie. To this we might add the observation that such intentions are quite uncommon in business negotiations. Even if I do not *expect* you to believe that $50,000 is my rockbottom price, I probably still *hope* or intend to deceive you into thinking that my minimum is higher than $45,000.

It seems that the timing and circumstances of a statement influence the way in which it is interpreted. Thus, the statement, "This is a nonnegotiable item," is likely to be interpreted quite differently when labor and management negotiators are stating their opening positions and when they are within one half hour of a strike deadline. Therefore, one and the same statement can constitute a lie in one set of circumstances and not a lie in another situation.

In *The Lectures on Ethics* Kant holds that a deliberate false statement does not constitute a lie unless the person who makes it has "expressly given" the other(s) to understand that he or she wishes to acquaint them with his or her private thoughts.[5] When I make a false statement to a thief about the location of my valuables, I am not lying because "the thief knows full well that I will not, if I can help it, tell him the truth and that he has no right to demand it of me.''[6] In Kant's view, false statements uttered in the course of business negotiations do not constitute lies except in the very unusual circumstance that one promises to tell the truth during the negotiating. Kant's definition of lying is very implausible. It rules out many common cases of lying. For example, suppose that a child standing in line for a ticket to see ''Deep Throat'' claims to be 18 when he or she is only 15. The child is lying in spite of the fact that the ticket taker was never told explicitly of an intention to speak frankly.

We can also read Kant's analysis of lying in either of the following ways: (a) a deliberate false statement does not constitute a lie unless the person to whom the statement is made has an expectation that the truth will be told, or (b) deliberate false statements are not lies unless the person who makes them believes that those to whom one speaks have the expectation that they will be told the truth. This analysis of lying has certain merits. For one thing, it allows us to say that deliberate false statements made in the context of games do not constitute lies. We have in mind such things as saying, ''I won't attack you on my next move if you don't attack me now,'' in the course of a game of ''Risk.'' A player in a game of Risk generally does not expect to be told the truth and does not believe that others expect the truth to be told either. One might liken business negotiations to a game of Risk and say that since no one expects to be told the truth during the game, no statements uttered during its course can constitute lies.[7]

There are two problems with this defense of the claim that business bluffing does not constitute lying. First, this argument only applies to cases in which the other party is familiar with standard business practices and does not expect to be told the truth. There would seem to be many cases where the other party is not familiar with standard negotiating practices (e.g., negotiating with children, naive individuals, immigrant laborers, or the mentally impaired).

Second, the proposed analysis of ''lying'' is implausible. Surely, it is possible to tell lies in situations in which one knows that the other parties do not expect to be told the truth. The deliberate falsehoods uttered by a criminal during the course of a trial are lies even if the criminal knows that the members of the court do not expect the truth to be told.

One important difference between speaking falsely in a game of Risk and speaking falsely to an incredulous jury is that in the latter case one has promised to tell the truth. The following definition recommends itself to those who would hold that, although intentional false statements made to a jury are lies, intentional false statements uttered during the course of a game of Risk are not lies:

> (4) A deliberate false statement constitutes a lie if and only if: (i) the speaker (writer, etc.) makes it with the intent that those to whom it is directed believe it, or foresees that they are likely to believe it, and (ii) either those to whom the statement is addressed expect to be told the truth, or the speaker has promised to tell them the truth. [Alternatively condition (ii) can be formulated as follows: (ii′) either the speaker knows or believes that those to whom the statement is addressed expect to be told the truth, or he/she has promised to tell them the truth.]

We are inclined to think that deliberate falsehoods uttered in the course of a game of Risk do constitute lies, and are thus inclined to prefer (3) to (4). This is a case about which people have conflicting intuitions; it cannot be a decisive reason for preferring (3) to (4) or vice versa. A more decisive consideration in favor of (3) over (4) is the following case. Suppose that a woman knows that her husband has been having an affair with another woman and asks him if this is so. He answers her by denying this. This would be a lie, even if she does not expect to be told the truth, and even if he believes that she does not expect to be told the truth.

There is, to the best of our knowledge, no plausible definition of lying that allows us to say that typical instances of bluffing in economic transactions do not involve lying. Therefore, we conclude provisionally that bluffing generally does involve lying. We should stress that it is only bluffing that involves making false statements that constitutes lying. One is not lying if one bluffs another by making the true statement "I want $50,000 for the house." Similarly, it is not a lie if one bluffs without making any statements, as in a game of poker, or overpricing (on a price tag) a product where bargaining is expected (e.g., a used car lot).

At this point, it would be useful to consider the relationship between lying and the broader concept of deception. Deception may be defined as intentionally causing another person to have false beliefs. (It is not clear whether preventing someone from having true beliefs should count as a case of deception.) As we have seen, lying always involves the intent to deceive others, or the expectation that they will be deceived as a result of what one says, or both. But one can lie without actually deceiving anyone. If you do not believe me when I lie and tell you that I have another person who wants to buy my product, then I have not succeeded in deceiving you about anything. It is also possible to deceive another person without telling a lie. For example, I am not lying when I deceive a thief into thinking that I am at home by installing an automatic timer to have my lights turned on in the evening.

It seems that one can often avoid lying in the course of a business negotiation simply by phrasing one's statements very carefully. In selling a house, instead of lying and saying that $50,000 is the lowest amount we will accept, I could avoid lying by making the following true, but equally deceptive statement: "My wife told me to tell you that $50,000 is the lowest price we will accept." (She may have told me this while we were plotting to deceive you.) It is questionable whether this is any less morally objectionable than lying. Most people prefer to deceive others by means of cleverly contrived true statements, rather than lies. Some who have strong scruples against lying see nothing wrong with such ruses. It is doubtful, however, whether lying is any worse than mere deception. Consider the following example. I want to deceive a potential thief into thinking that I will be at home in the late afternoon. I have the choice between (i) leaving my lights on, and (ii) leaving a note on my door that says "I will be home at 5 PM." Surely this choice is morally indifferent. The fact that (ii) is an act of lying and (i) is not does not itself constitute a reason for thinking that (i) is morally preferable to (ii).[8]

The Morality and Economics of Lying

Common sense holds that lying is a matter of moral significance and that lying is *prima facie* wrong, or wrong everything else being equal. This can also be put by saying that there is a presumption against lying, and that lying requires some special justification in order to be considered permissible. Common sense also holds that lying is not always wrong, it can sometimes be justified.[9] Almost no one would agree with Kant's view that it is wrong to lie even if doing so is necessary to protect the lives of innocent people. According to Kant it would be wrong to lie to a potential murderer concerning the whereabouts of his intended victim.[10] Common sense also seems to hold that there is a presumption against simple deception.

Assuming the correctness of this view about the morality of lying and deception, and assuming that we are correct in saying that bluffing involves lying, it follows that bluffing and other deceptive business practices require some sort of special justification in order to be considered permissible. Businessmen frequently defend bluffing and other deceptive practices on the grounds that they are profitable or "economically necessary." Such acts are also defended on the grounds that they are standard practice in economic transactions. Are such grounds sufficient to justify lying and deception given our present assumptions? There are four separate questions that we must consider here:

(1) Are lying and deception ever profitable or "economically necessary?"

(2) If it could be shown that they are either profitable or "economically necessary" would that be enough to justify them?

(3) Are lying and deception standard practice in economic transactions?

(4) Supposing that they are standard practice would that justify them?

(1) There are those who hold that lying and deception are never profitable or "economically necessary." In their view, honesty is always the best policy. One incentive for telling the truth is the law, but here we are referring to lying or bluffing that is not illegal or for which the penalty or risk of being caught is not great enough to discourage the action.

Those who hold that honesty is always in one's economic self-interest would argue that economic transactions are built on trust and to violate that trust would discourage an individual or organization from entering into further transactions with the lying party for fear of being lied to again. Thus, some mutually beneficial transactions may be foregone for lack of trust. Moreover, those adhering to this position would argue that word of the deceitful practices would spread throughout the marketplace and others would also avoid doing business with the liar. Thus, although some short run profit might accrue from lying, in the long run it would be unprofitable. If this argument were sound, we would have a non-issue. Lying, like inefficiency, would be a question of bad management that would be in one's own best interest to eliminate.

Unfortunately there are some anomalies in the marketplace that prevent the system from operating in a perfectly smooth manner. The very existence of bluffing and lying in the first place suggests that the economist's assumption of perfect (or near pefect) market information is not valid. Some transactions, such as someone's buying or selling a house, are one-shot deals with little or no chance of repeat business. Thus, there is no experience on which to base an assessment of the seller's honesty and no incentive to build trust for future transactions. Even where a business is involved in an ongoing operation, information flows are such that a large number of people can be duped before others hear about it (e.g., selling Florida swampland or Arizona desertland sight unseen in other parts of the country). Other bluffs and lies are difficult if not impossible to prove. If a union negotiator wins a concession from management on the grounds that the union would not ratify the contract without it—even though the negotiator has reason to believe this is untrue—it would be extremely difficult for management to later prove that ratification could have been achieved without the provision. By the same token, some product claims such as the salesman's contention that "this is the best_____ on the market" are inherently subjective. When the competing products are of similar quality, it is difficult to prove such statements untrue even if the person making the statement believes them to be untrue. Another exception to the assumption of perfect information flows is the confusion brought on by the increasing technological complexity of goods and services. In fact, a product information industry in the form of publications like *Consumers' Guide, Consumer Reports, Money,* and *Changing Times* has arisen to provide, for a price, the kind of product information that economic theory assumes consumers have to begin with.

These arguments suggest that not only are the commonly cited disincentives to bluffing and lying often ineffective, but that there are some distinct financial incentives for these activities. If you can convince consumers that your product is better than it really is, you will have a better chance of selling them that product and/or you may be able to charge them a higher price than they would otherwise be willing to pay. It is also obvious that in a negotiating setting there are financial rewards for successful lies and bluffs. If you can conceal your actual minimal acceptable position, you may be able to achieve a more desirable settlement. By the same token, learning your negotiating opponent's true position will enable you to press towards that minimal acceptable position. This is, of course, why such intrigues as hiding microphones in the opposing negotiating team's private quarters or hiring informants are undertaken in negotiations—they produce valuable information.

(2) An individual cannot justify lying simply on the grounds that it is in his or her own self-interest to lie. For it is not always morally permissible to do what is in one's own self-interest. I would not be justified in killing you or falsely accusing you of a crime in order to get your job, even if doing so would be to my advantage. Similarly, a businessperson cannot justify lying or deception *simply* on the grounds that they are advantageous (profitable) to the business. This point can be strengthened if we remember that any advantages one gains as a result of bluffing are usually counterbalanced by corresponding disadvantages on the part of others. If I succeed in getting a higher price by bluffing when I sell my house, then there must be someone else who is paying more than might otherwise have been the case.

"Economic necessity" is a stronger justification for lying than mere profitability. Suppose that it is necessary for a businessperson to engage in lying or deception in order to insure the survival of his or her firm. Many would not object to a person stealing food to prevent starvation of his or her family. It would seem that lying in such a situation to get money to buy food or to continue employing workers so that *they* can buy food would be equally justifiable. This case would best be described as a conflict of duties—a conflict between the duty to be honest and the duty to promote the welfare of those for/to whom one is "responsible" (one's children, one's employees, or the stockholders whose money one manages.) However, it is extremely unlikely that bankruptcy would result in the death or starvation of anyone in a society that has unemployment compensation, welfare payments, food stamps, charitable organizations, and even opportunities for begging. The consequences of refraining from lying in transactions might still be very unfavorable indeed, involving, for example, the bankruptcy of a firm, loss of investment, unemployment, and personal suffering associated with this. But a firm that needs to practice lying or deception in order to continue in existence is of doubtful value to society. Perhaps the labor, capital, and raw materials that it uses could be put to better use elsewhere. At least in a free market situation, the interests of economic efficiency would be best served if such firms were to go out of business. An apparent exception to this economic efficiency argument would be a situation in which a firm was pushed to the edge of bankruptcy by the lies of competitors or others.

It seems probable to us that the long-term consequences of the bankruptcy of a firm that needs to lie in order to continue in existence would be better or no worse than those of its continuing to exist. Clearly, on this assumption a businessperson cannot justify lying or deception in order to avoid bankruptcy. Because, in that case, the strong presumption against lying and deception is not counterbalanced by any other considerations (the long-term consequences of the firm's going out of business are no worse than those of its continuing to exist). Suppose, however, that the immediate bad consequences of bankruptcy would not be offset by any long-term benefits. In that case it is no longer clear that it would be wrong for a company to resort to lying and deception out of "economic necessity." One can be justified in lying or deceiving to save individuals from harms far less serious than death. I can be justified in lying about the gender of my friend's roommate to a relative or nosey boss in order to protect that friend from embarrassment or from being fired. If the degree of harm prevented by lying or deception were the only relevant factor (and if bankruptcy would not have any significant long-term benefits), then it would seem that a businessperson could easily justify lying and deceiving in order to protect those associated with the business from the harm that would result from the bankruptcy of the firm. There is, however, another relevant factor that clouds the issue. In the case of lying about the private affairs of one's friends, one is lying to others about matters concerning which they have no right to know. Our present analogy warrants lying and deception for the sake of economic survival only in cases in which the persons being lied to or deceived have no right to the information in question. Among other things, this rules out deceiving customers about dangerous defects in one's products—they have a right to this information—but it does not rule out lying to someone or deceiving him/her about one's minimal bargaining position.

We have argued that personal (or corporate) profit is no justification for lying in business transactions and that lying out of "economic necessity" is also morally objectionable in many cases. But what about lying in order to benefit the party being lied to? There are certainly many self-serving claims to this effect. Some have argued that individuals derive greater satisfaction from a product or service if they can be convinced that it is better than is actually the case. On the other hand, an advertising executive made the argument in the recent FTC hearings on children's advertising that the disappointment children experience when a product fails to meet their commercial-inflated expectations is beneficial because it helps them develop a healthy skepticism. These arguments are not convincing and they appear to be smoke screens for actions taken out of self-interest. Deceptive advertising is almost invariably engaged in for reasons of self-interest even though it is conceivable that consumers might benefit from it. For example, deceptive or puffed up advertising claims may cause one to purchase a product that is of genuine benefit. One example where lying or deception actually appears to be engaged in for the consumer's interest is the fairly common and often successful practice of doctors prescribing placebos to eliminate patients' real or imagined illnesses. Having been convinced that the inert substance is effective, the patient may actually find relief from his or her symptoms.

This practice presumably differs in motive from the activities of so-called medical quacks and rip-off artists who sell such things as copper bracelets, or laetrile that may also have placebo effects. It would seem that the doctor has nothing to gain by administering the placebo as opposed to an active drug that might have some negative side effects. The purveyors of copper bracelets and laetrile have much to gain by having people purchase their products rather than undergo standard medical procedures. In considering the morality of such activities, it is worth noting that there is probably a greater risk to the patient associated with foregoing all standard medical treatment to undergo laetrile therapy than there is associated with receiving a placebo while under a doctor's care. Given this, many would feel that the benefits of a doctor deceiving his or her patients into better health constitute a reasonable exception to his *prima facie* duty to tell the truth. Before condoning this practice, however, one should consider the possible adverse effects on the patient of discovering the deception and also the cumulative impact on society of this practice of deception by the medical profession.[11] Thus other avenues of treatment should be explored before placebo giving is considered.

Although lying and deception can sometimes be justified by reference to the interests of those being lied to or deceived, such cases are very atypical in business situations. As was argued earlier, successful bluffing almost invariably harms the other party in business negotiations. The net effect of a successful bluff is that the bluffed party pays more or receives less than would otherwise be the case.

(3) and (4) Lying and deception are very common (if not generally accepted or condoned) in business transactions. Bluffing and other deceptive practices are especially common in economic negotiations and bluffing, at least, is generally thought to be an acceptable practice. Does this fact in any way justify bluffing? We think not. The mere fact that something is standard practice or generally accepted is not enough to justify it. Standard practice and popular opinion can be in error. Such things as slavery were once standard practice and generally accepted. But they are and *were* morally wrong. However, the fact that bluffing is common can justify it indirectly. If one is involved in a negotiation, it is very probable that the other parties with whom one is dealing are themselves bluffing. It seems plausible to say that the presumption against lying and deception holds *only* when the other parties with whom one is dealing are not themselves lying or attempting to deceive one. Given this, there is no presumption against bluffing or deceiving someone who is attempting to bluff or deceive you. It should be stressed again that the prevalence of bluffing *per se* is no justification for bluffing oneself. The justification for bluffing a particular individual in a particular situation derives from the fact that he or she is attempting to deceive you on that occasion. The fact that bluffing is so common only means that there are many situations in which it can be justified on these grounds. In fact, there is such a strong presumption for thinking that the other parties will bluff or lie in negotiating settings that one is justified in presuming that they are lying or bluffing in the absence of any special reasons to the contrary, e.g., one's dealing with an unusually naive or scrupulous person.

A further ground on which lying or deceiving in bargaining situations is sometimes held to be justifiable is the claim that the other parties do not have *a right to know* one's true bargaining position. It is true that the other party does not have a right to know one's position, i.e, it would not be wrong for one to refuse to reveal it to that person. But this is not to say that it is permissible to lie or deceive him or her. You have no right to know where I was born, but it would be *prima facie* wrong for me to lie to you about the place of my birth. So, lying and deception in bargaining situations cannot be justified simply on the grounds that the other parties have no right to know one's true position. This is not to deny that, other things being equal, it is much worse to lie or deceive about a matter concerning which the other parties have a right to know than one about which they have no right to know.

As we have argued, there appears to be a personal economic incentive in many cases for lying, bluffing, and deception. We, therefore, cannot rely solely on the marketplace to eliminate these activities. Thus, we now consider what sorts of guidelines we can offer the individual buying or selling goods or services who is concerned about the morality of lying and bluffing, but is also concerned about furthering his or her own economic welfare.

Bluffing is an attempt to distort one's true bargaining position. However, one could gain some negotiating advantage by simply concealing that position. Thus, if someone asks the minimum price you would accept for your house, you could say, "I want $50,000 for this house," without lying or deceiving anyone and without revealing the minimum offer you would accept. Although this is clearly not as strong a position as "I won't accept a penny less than $50,000 for this house," if the prospective buyer felt that it was worth $50,000 and that someone else might buy it at that price, he or she might be willing to pay that amount. Whether or not this happens, of course, depends on the relative bargaining positions of the negotiators. In addition to this "truthful negotiating" approach, we make the following recommendations for personal and institutional responses to the problems involved in lying, bluffing, and deception in business:

(1) Explicit consideration by negotiators and all persons engaged in economic transactions of the ethical issues involved in their activities. Even if there is no general consensus about the moral status of an action, a well-thought-out action would seem to be preferable to one made without any attention to moral issues.

(2) Focusing on the morality of bluffing and lying not only in negotiations courses such as Professor Raiffa's, but also as part of business ethics courses.

(3) Strict legal definitions of the illegal deceptive practices in negotiations. If the market system cannot provide adequate incentive for truth-telling, perhaps the legal system can.

(4) Utilizing ethical advisors for opinions on the morality of certain economic transactions.[12] A resident philosopher would, of course, be practical only for large corporations and unions, but even individuals could seek counsel from trusted friends or clergy.

(5) Further analysis of the moral status of lying and deception in economic transactions. The following issues seem particularly important:

(a) To what extent is the general presumption against lying and deception overridden when the other parties are lying or deceiving themselves, or when they had no right to the information in question?

(b) The definition of deception. Differences between lying and deception. Is it permissible to avoid lying by making true but deceptive statements?

Sisella Bok suggests two other approaches for dealing with lying:

(6) The development of financial and other incentives for honesty in economic and political transactions and organizations.[13]

(7) She suggests the need for public participation in deliberations on the issue of lying in business and business ethics in general. For example, public representatives on regulatory commissions considering these issues or on business panels seeking to develop codes of ethics would provide for the public scrutiny which she feels is necessary for forthright consideration of ethical issues.

Although these steps would not eliminate the uncertainty about the morality of all the activities we have discussed, they would provide an improvement over decision-making systems that ignore ethical issues.

Acknowledgment

We are indebted to Thomas Beauchamp for comments on a previous version of this paper.

Notes and References

[1]William M. Bulkeley, "To Some at Harvard Telling Lies Becomes a Matter of Course," *Wall Street Journal* (January 15, 1979). 1, 37.

[2]Arnold Isenberg defines a lie as follows: "A lie is a statement made by one who does not believe it with the intention that someone else shall be led to believe it. This definition leaves open the possibility that a person could be lying, even though what he says is true . . .", "Conditions for Lying," in *Ethical Theory and Business,* Tom Beauchamp and Norman Bowie, eds. (New Jersey: Prentice-Hall, 1979), p. 466.

[3]Roderick Chisholm and Thomas Feehan, "The Intent to Deceive," *Journal of Philosophy* (March 1977), 152.

[4]Fredrick Seigler considers this kind of example in "Lying," *American Philosophical Quarterly* (April 1966). But he argues that it does not count against the view that a necessary condition of a statement's being a lie is that it is intended to deceive someone. The example only shows that it is not necessary that the liar personally intend to deceive the others. But it does not count against the view that the lie must be intended *by someone* to deceive others. For, in our present example, *the criminal intends* that the witness's statements deceive others. Siegler is correct. However, a slight modification of the present example generates a counter-example to his claim that a lie must be intended by someone or other to deceive. Suppose that a witness makes a deliberate false statement, *x*, for fear of being killed by the friends of the accused. The witness is lying even if the accused's friends believes that *x* is true, in which case neither they nor anyone else intend that the witness's statements deceive the jury.

[5]Immanuel Kant, *Lectures on Ethics,* trans. Louis Infield (New York: Harper and Row, 1963), p. 228. This book is based on notes from Kant's students during the period 1775–1781. The analysis of lying offered here differs from the one presented in Kant's later and more well-known work, "On the Supposed Right to Tell Lies from Benevolent Motives" (1797), in Barauch Brody, ed., *Moral Rules and Particular Circumstances* (New Jersey: Prentice-Hall, 1970). There he says that any intentional false statement is a lie (p. 32). Kant also gives a different account of the morality of lying in these two works. His well-known absolute prohibition against lying is set forth only in the latter work.

[6]Kant, *Lectures on Ethics,* p. 227.

[7]Cf., Albert Carr, "Is Business Bluffing Ethical," in *Ethical Issues in Business,* Thomas Donaldson and Patricia Werhane, eds. (New Jersey, Prentice-Hall, 1979).

> I quoted Henry Taylor, the British statesman who pointed out that "falsehood ceases to be falsehood when it is understood on all sides that the truth is not to be spoken"—an exact description on bluffing in poker, diplomacy, and business. (p. 46.)

[8]We owe this example to Bernard Gert.

[9]The classic statement of this view is included in Chapter II of Sir William David Ross' *The Right and the Good,* (Oxford, Clarendon Press, 1930).

[10]"On the Supposed Right to Tell Lies from Benevolent Motives," pp. 32 and 33.

[11]See Sissela Bok, *Lying: Moral Choice in Public and Private Life.* (New York: Vintage Books, 1978).

[12]See John T. Steiner "The Prospect of Ethical Advisors for Business Corporations," *Business and Society,* Spring, 1976, pp. 5–10.

[13]See Jerry R. Green and Jean-Jacques Laffont, *Incentives in Public Decision Making* (Amsterdam: North-Holland Publishing Co., 1978); and William Vickery, "Counterspeculation, Auctions, and Cooperative Sealed Tenders," *Journal of Finance* (March 1961): 8–37, cited in Bok, p. 333.

The Social Business of Business

Richard T. De George

Imagine a slaveholder trying to be fair in settling a dispute between two slaves, trying to be just in distributing the slop made available to the slaves as food, or in making sure that they are all whipped equally, or at least equally for equal offenses. There is certainly a sense in which the slaveholder is trying to be fair, just, or moral within the slave-owning system. Yet most Americans would readily admit that slavery is immoral. The slaveholder who tries to be moral within the system of slavery may have some grounds for feeling superior to the slaveholder who cares not at all about the slaves, who beats them arbitrarily, who starves them out of malice, or who settles disputes between them by whim. But surely a discussion of the moral way to treat one's slaves, though it might do some good, has an odd ring to it. For it deals with the morality of practices within a system that itself should be morally evaluated—and if morally evaluated by our standards is morally condemned.

Are we in a similar situation? As we congratulate ourselves on our moral perspicacity in condemning bribery, or on our moral courage in raising questions about the practices of multinational corporations, or as we speak of a just wage, are we comparable to the slaveholder? Are we attacking moral problems within a system when we should actually be attacking the system?

My answer, which I have argued for elsewhere,[1] is that though the system has many elements that are immoral, it is not inherently immoral. And even if it were in some sense immoral in itself, so far as I can see there is no other morally superior system waiting to be adopted. We have no panacea waiting in the wings. We have no alternative system waiting for a revolution to see it actualized. The goal of a proposed alternate system to replace ours is the proposal of pie in the sky. For even if there were such a system waiting for us to grasp it, if by "we" we mean the American people, it is abundantly clear that we are not about to have a revolution, that no one has a mandate to lead one, and that we are not about to buy communism or even self-management socialism.

To claim this, however, is not to argue for the status quo, nor to argue for no change in the system. For my thesis is that though there is no mandate for revolution, there is a mandate for business to change some of its goals, methods, and organization.

There is a cliche that states: the business of business is business. The cliche is also a tautology. This means of course that it is true. But some tautologies are interesting and some are not. The question here is what is business and who decides? The business of business is not government or social welfare. But business in socialist countries, despite similarities with business in the United States, is also notably different from it. In Japan the business of business includes guaranteeing continual employment for a firm's employees. This is not—or not yet—the business of American businesses. The point, however, should be clear. Despite the least common denominator that all these systems share of providing goods and services and of distributing them, the business of business varies from society to society and from social system to social system. What business is, how wide its scope and responsibility are or ought to be, and how it is to pursue its ends are all questions that are determined differently in different societies. In a democracy the people should decide what the business of business is, just as they decide what the business of government is.

In the United States the people's mandate to business is different now from what it was twenty years ago and incomparably different from what it was 200 years ago when the Constitution was adopted. Businesses are now coming under increasing demands from numerous quarters to lower prices, increase profits, preserve the environment, raise wages, bolster job security, disclose more information about their operations, engage in social as well as financial accounting, and help solve the problems of poverty, discrimination, and urban blight.

The reaction of business for the most part has been to proceed with business as usual until forced by legislation to do otherwise. There are of course exceptions. But on the whole business has felt that the attacks upon it are unjustifiable, that it is being charged with correcting social ills that are not of its making, that it is being asked to undertake tasks that are not part of its goal, that its freedom is being illegitimately threatened and its autonomy unjustly undermined.

Faced with conflicting demands, many of which seem counter to their interests, many corporations have not known how to respond. Not knowing, many of them have evidently decided to ride out the storm, doing what they are forced to do by legislation, but hoping that such things as consumerism and demands for social accounting will eventually go away. A few corporations have indicated that they would like to comply with the new demands being placed on them, if only they knew how. But, they complain, the demands are vague, sometimes at odds with one another, and no one except government spells them out clearly. Even fewer corporations have attempted to reply either by taking positive action to preempt harsh legislation or by mounting public counterattacks explaining and defending their view of the situation to all interested parties.[2]

I do not believe that the lack of effective and positive response comes from bad faith. But standing in the way of an effective response is an outdated image of what the corporation is, and a corresponding lack of organizational structure corresponding to the new demands. If I am correct, what is needed is both a different image of the corporation and corresponding organizational changes. Although my suggestions might seem like one more demand being placed on business, I suggest that the new model, since it corresponds to social realities, will help businesses both understand and respond to social demands more realistically. I do not believe that I or anyone else outside of a corporation can or should supply a blueprint for handling competing social demands, for they differ from corporation to corporation, industry to industry, and place to place. What is appropriate for a small business is different from what is appropriate for a multinational conglomerate. But businesses of any size need procedures and structures that will bring legitimate demands to the attention of those who must face and weigh them. I shall argue that these should flow from the changing mandate society is in the process of giving to business.

The Changing Mandate to Business

The private business corporation as found in the United States has a history that explains in large part its present function and structure. Ultimately, the business corporation receives its legitimacy from society and must be flexible enough to change as society changes. There is no one kind of business organization that is sacrosanct or necessary. The nationalization of industries in many countries, the socialization of industries in other countries, the threat to private industry in countries facing the possibility of Eurocommunism, and the increasing restrictions on American free enterprise clearly show that industries can be operated in a great many different ways. Business executives are mistaken if they feel they are immune from social takeovers, influences, or demands. If private industries do not meet social demands, the alternative of governmental control or even of governmental ownership remains a possibility that has become an actuality in many countries.

For many years the mandate that American society gave to business was to grow, to increase profits, to produce consumer goods at the lowest possible price, and to raise the American standard of living. Business responded to the mandate. The mandate was relatively simple and no one could complain if business responded simplistically. For a while it seemed almost true that what was good for General Motors was good for America. But as society became more complicated, as all of the pieces of the social complex became more interdependent, the freedom of business to pursue its own ends as it saw best has slowly been curtailed. Such things as air and water, which were thought to be free, have come to be seen as damageable and limited.

Business for the most part has continued to see itself in its former mold, and it frequently continues to pursue its historical goals. There are, however, increasing signs that the old American mandate to business had changed. The change has come gradually and has not been sufficiently articulated. But the change is now significant enough to constitute a revised mandate, and it is one that business can continue to ignore only at its own peril.

To speak of a changed mandate for business may sound like a personal intuition, or simply one person's point of view. I find it objectively, however, not only in movements such as consumerism and in public outcries over bribes and windfall profits, but in legislation. Business has opposed legislation dealing with environmental protection, worker safety, consumer protection, social welfare, affirmative action, truth in lending, fair packaging and labeling, truth in advertizing, child labor, workmen's compensation, minimum wages, pension reform, and so on. Legislation has been passed in all these areas over the objections of business. Why has business been opposed to such legislation—legislation that in most instances seems progressive, socially desirable, and in the public good? In each case business decried the encroachment of government and claimed that it was protecting profits. Yet all this legislation has not prevented business from prospering and from making profits. The legislation has expressed social demands and it embodies a view of business that, when taken as a whole, is clearly different from the eighteenth century view of business found in the writings of John Locke or in the Constitution, and different from the simplistic mandate given to business in an earlier time. The fact that business has responded negatively to each such piece of legislation shows that it is less responsive to popular demands than many people think it should be. As a result it has been labeled by the general public as self-seeking, narrowly interested, and socially blind. Even businesspeople themselves have a poor image of themselves.[3] The fact that business has prospered despite such legislation shows that it is more resilient and more able to face social demands than many of its leaders believe or would have us believe. Given the situation I described above, I suggest that business will be forced to do more and more in the line of social concerns, and that it should be so structured as to respond positively before it is forced to respond under legislation. The creative genius of American business, if put to the test, can undoubtedly come up with better solutions to many problems than those forced upon it in procrustean legislation.

The Traditional Corporate Model

I have suggested that the negative response of business is in part a result of business's view of itself. The American business corporation developed in a certain way. As it developed it was studied and described by sociologists and organizational theorists.[4] In an attempt to be value free in their approach, they described the business corporation as they found it. Their descriptions have in turn become the model taught to business students and rising young managers. These managers have believed what they were taught and have come to hold that the descriptions correspond to what the corporations *should* be. This is a mistake. It is a mistake that can and should be corrected. And it is a mistake that, if corrected, can have significant results in freeing the corporation for organizational change.

According to the traditional model of an organization, a corporation is a legal entity founded for certain limited aims—profit, production, provision of services, and so on. It is organized to fulfill these limited goals. The corporation employs people to carry out their assigned tasks. Employees of the corporation are paid to act not as individuals in a private capacity, but as impersonal agents of an organization. Each person has a function that is to be carried out in accordance with the goals of the corporation, the policies set by the managers, and the procedures developed to achieve the organization's ends. Each person occupies a position and within that position is replaceable by other people, any number of whom can fulfill the same function.

On this traditional model the organization is neither an individual nor a moral entity. To speak of it in moral terms is, therefore, according to the theorists, inappropriate. Since its employees act not as individuals, but as impersonal agents functioning within and for the organization, they should not let their personal moral notions supercede the ends of the organization; nor are they personally morally responsible for the actions of the organization. Moral responsibility is therefore inappropriately applied to either corporations or to the employees of corporations when acting as corporate agents.[5] As legal creations corporations may be legally restrained and they may have legal responsibility. But morality is not part of what they can logically be held responsible for.

It follows from this traditional view that morality is improperly thought to be part of a corporation's concern. It is only because people are confused on this issue that some of those outside a corporation judge it from a moral point of view, or accuse it of failing to satisfy moral demands, or expect it to consider some set of moral values that it must implement as if it were a moral being. Laws, on this view, must be complied with; but under the democratic process, its defenders argue, when bills are not in the interest of the corporation, they should be fought. And moral demands, since they are the result of muddled thinking and are improper considerations, should rightly be ignored by corporations.

Corporations are freely formed to achieve certain ends, the theory continues. They are formed by those willing to risk their capital. Individuals are guaranteed their freedom under law, and so are free to form groups that can achieve ends they are not able to achieve individually. These groups, whether informal or formalized as corporations, should be free to pursue their ends, just as individuals are free to do so. They may be constrained by law. But their freedom should be guaranteed, just as the freedom of individuals who form them are. If they hire others to work for them, the workers do so freely and agree to work for certain wages. No one is forced to work at a given job. No one is forced to buy a certain product or service. Since all the transactions are freely entered into, there is no coercion, and the corporation should be allowed to pursue its ends as it sees fit and best. This is part of the American heritage and part of what it means to live in a free country.

Now this theory and description of organizations, and especially of corporations, seems to be one that is accepted by many people. It represents the views of many workers, managers, and owners of corporations. Given this view, those who hold it are understandably annoyed, perplexed, and bewildered by those who wish to evaluate corporations and other organizations from a moral point of view, who wish to impose ends and goals and practices on them that have little to do with their freely chosen and designated goals and ends, and that, e.g., in the case of profits, may threaten those ends. By what right, they ask, does society or government or those not inside the organization or corporation impose extraneous demands on it, attempt to force moral norms on it, and constrain it in a variety of ways?

The annoyance, perplexity, and bewilderment, I suggest, arise in part because the organizational model and theory I have described are not, and have never been, value free. They have been parts of an ideology that may have once dominated the ordinary consciousness of most people in our society. But it is an ideology that is now in the process of erosion and it is a description that is too narrow a view of corporations in a world in which all people are increasingly interdependent, in which the myth of radical individualism no longer holds sway, and in which the actions of corporations impinge quickly and seriously on those outside of them in a way that is no longer passively tolerated.

The Defects of the Model

Corporations are not the free creations of individuals. They are legally constituted social entities. They exist by social allowance. They are allowed if and when they fulfill social needs, and they are legitimately restrained by social demands. There are three ideological components of the model organizational theory that stand in the way of a corporation's adequately seeing and responding to changing societal demands. The first is that the corporation, as a free organization, by right sets its own ends without interference. The second is that the hierarchical system of authority in a corporation is justified in terms of ownership and efficiency. The third is that because the corporation is an impersonal entity, it and its agents have no moral responsibility. All three need rethinking.

The traditional organizational view is, from a moral point of view, outrageous. Since it is a category mistake on this view to apply moral language to an organization, the organization has moral immunity. Thus, though murder by an individual is morally condemned, Murder Inc. cannot be faulted for pursuing its goal; nor can Advertisers Inc. for lying; nor Shoddy Inc. for producing dangerous tools; nor Hitler's SS for exterminating Jews. They may, of course, be legally restrained; but they are morally immune. The organizational theory that describes and defends this view is clearly an ideology. It is accepted by many who defend the status quo and it provides a convenient excuse for corporate immorality.

Corporations are of course *legal* and not *human* persons. They have no feelings, they have no conscience, they are not moral agents. As the corporate shield falls away, we see clearly that a corporation is simply a piece of paper. At best corporations are structures. They are, however, structures filled by people. It is people who adopt goals; it is people who make decisions; it is people who implement policy. It is also people who are moral agents. The basic difficulty is not in finding some way to understand organizations as moral agents so they can be held morally responsible. Rather, it is to overcome the reified view of organizations so that we see they cannot act without people, and it is people who are morally responsible. The people in an organization are responsible for what an organization does. If in the name of the organization they lie, cheat, kill, maim, steal, they are morally responsible for these actions. The myth that they are agents of an organization, that their actions are to be attributed to the organization and not to them, and that the individual is not responsible is a myth that should be shattered, a mistake that should be corrected, a dogma that should be rejected.

People cannot be rid of their moral or legal responsibility simply by disclaiming it, or by seeking anonymity in corporate decisions, or by pretending decisions are impersonal or made by a corporation or organization that is not a moral being and so not morally accountable. Though the corporation is not a moral agent, the people in a corporation are moral agents and both the actions done by corporations and the harm done to individuals can and should be morally evaluated.

If taking bribes is immoral, then those in a corporation who set the policy of taking bribes as part of their way of doing business are morally culpable as well as those who take bribes. But who sets policy? Is it not most often a joint decision to which many people contribute? And since it is a corporate decision, does not the guilt of each become really insignificant? This is what organizational theory suggests. But take a clear case and see if the argument stands. Suppose that a company cannot make it competitively unless the genius running the competing firm is eliminated. A fatal accident is arranged and the world has one less genius. Is there any doubt that morally all those involved are guilty of murder? To claim one is less guilty simply because he or she voted for, rather than actually carried out, the killing, is from a moral point of view indefensible. The same is true of theft, lying, and any other immoral action. Individuals do not become robots and do not stop being moral beings when they occupy places in organizations. Morality enters the corporation with individuals or it does not enter at all. It is they who are morally responsible, or there is no moral responsibility involved in what organizations do. The Nuremburg trials and the trial of Cally at My Lai represented instances in which organizational responsibility was personally placed. The GM scandals and other similar cases are straws in the wind indicating that corporate responsibility is increasingly being ascribed to individuals.

The ideal organizational model described by social scientists might be no more than a language game that they or corporate executives wishing to avoid moral responsibility would like to play. I suggest that the rules of the game are not theirs to make. Organizations derive their legitimacy, as do governments, from the people. They are, in fact, structures within which people act to achieve their ends. But they should be restricted, as individuals are restricted, in their actions if their deeds have adverse effects on others.

Now my claim is that in a variety of ways—through court decisions, through legislation, through popular reaction to the Gulf Oil scandals, to the DC-10 cases, to Watergate, and so on—there is a new, still incompletely articulated mandate to the corporation emerging. We do not disagree on what is moral in these cases. The mandate is to change corporate structure so as to prevent or minimize such occurrences.

A New Corporate Model

What I want to do now is sketch a view of the corporation from which the new social demands can be seen to flow logically and with some consistency. To reconceive the corporaton is, of course, not enough. It must be reorganized. What are presently external demands should be internalized and seen as properly part of the cost of doing business.

To speak of "humanistic capitalism" may be a contradiction in terms. But if humanism involves concern for human beings, the new view of business is that those who run it should not only be moral, but should also be concerned for the people who work in business and for those affected by it. The time when people could be considered simply replaceable parts on the assembly line is over. Safety, health insurance, retirement funds, educational and retraining opportunities are already considered entitlements. A corporation affects the lives of people by its selection of a factory site, by its decisions on how and whom to hire, and by its determination of when and how to relocate its plants. It can no longer make such decisions by the simple fiat of those at the top.

The authority for a corporation to carry on its business comes from the people. The doctrines of democracy and of consent of the governed are cornerstones of the American political system. As business and government have become more and more intermingled, the need for consent and democratization in business has increased. As workers through their pension plans and insurance policies own more and more company shares, the line between owner and worker blurs. Managers do not own the corporation and are as much hired hands as the janitors. Workers have not asked to run the factories, but they have been fighting for humanistic treatment. There are limits to the obedience that can be demanded of them and they have rights that they do not sell with their labor.

The new basis for corporate development may appropriately be called democratic humanism. It is democratic in the sense that the corporation must consider the rights, demands, and views of the worker, the consumer, and the general public—as well as the shareholder—in making decisions. It is not owned by those at the top. Its authority comes from below. It is humanistic in the sense that the good of all the people affected by its actions must be considered. Social accounting as well as financial accounting is now required.[6]

The upshot has been increasing demands made on business. Basic to all of them are two: responsibility and responsiveness. The key to handling these two is the key to handling conflicting demands made upon business.

If the basic demands on business are for responsible action and responsiveness, what is required is not some general code of business ethics similar to the code of ethics for lawyers, or accountants, or doctors. Business is not a homogeneous enterprise in the way that law, accounting, and medicine are. Any code general enough to cover all business will end up being something like the Ten Commandments. Surely we do not have to be reminded that lying is wrong, that theft is wrong, or that murder and maiming are wrong. Particular applications of these in particular circumstances are what vary and what various groups might draw up for themselves. Codes make sense, however, only when they apply to a rather homogeneous set of activities. They are respected by those outside the profession only when they are not self-serving, when they safeguard the public, when they are reliably invoked and enforced, and when those in violation of them are effectively and appropriately sanctioned. When a profession or group does not police itself, it can expect to be policed by others.

Whether there are corporations or industries that feel the need for developing and enforcing ethical codes for themselves is a decision to be made by members of those groups. Some engineers, for instance, are formulating ethical codes in their professional organizations. These organizations are also attempting to bring their weight to bear in defense of members of the profession who find that they must raise moral issues within their corporations of such a nature that if ignored could lead to whistle blowing. Whether secretaries through their organizations wish to act similarly, whether particular types of businesses wish to attempt to do so, whether better business bureaus wish to draw up codes of various types, is necessarily up to those involved. For codes by their very nature are self-developed, self-adopted, and self-enforced. The existence of such codes and the acceptance by any significant number in a group form the basis for pressure for general adoption and enforcement.

The solution to handling competing societal demands, however, is not to be found in ethical codes, nor is it to be found in any other substantive set of guidelines. For sometimes the demands of employees will carry greater weight than those of stockholders; sometimes the opposite will be the case; and sometimes both will have to give way to environmental demands. What is different now from the past is not that there are conflicting demands made on business, but that some of the demands cannot easily be stated in cost accounting terms. It is because the social audit frequently does not lend itself to any sort of precise measurement that it is so difficult to deal with.

The solution to handling conflicting societal demands is to be found, I suggest, not in substantive guidelines, but in procedural guidelines. This approach leaves the ultimate decisions where they belong—in the hands of management. But it holds management responsible both for them and for mistakes, making management more vulnerable than under the traditional model of a corporation. More importantly, it requires restructuring of the corporate organization itself. Codes and substantive guidelines are superimposed on an organization that continues to function in most respects as it did before. Procedural guidelines call for internal modifications, so that in some ways the corporation no longer functions as it did before.

By what right, it might be asked, can anyone require such changes? The reply, I think, is that organizational changes are the only ones that can enable the corporation to handle the many other demands placed upon it and survive in anything like its present form. For to stick to the traditional model, to refuse to consider the social dimensions of a corporation's activities, to take positive social action only when and as forced, is to ask for increasingly harsh and restrictive legislation that will bury the corporation as it was and eventually replace it with governmental control, governmental planning, and finally governmental ownership.

The Organizational Changes

The organizational changes required to handle conflicting societal demands grow out of the need for organizations to be both responsible and responsive.

In the traditional corporate model, legal responsibility was attached to the corporation as such, and not to individuals. In the new model both legal and moral responsibility must be assigned to individuals. Courts have recently been holding members of the board responsible for actions of a corporation and the actions it takes in the name of the corporation. Both the courts and the people have come to see that corporations are people. The first change required, therefore, is:

1. Assign responsibility, both legal and moral, at every level and see that the responsibility is assumed.

Clearly, if a board is to be held responsible, it cannot simply be a rubber stamp. The board members must be informed and concerned. They are responsible to the government for fulfilling their legal obligations, to the public for not violating the public's moral and legal rights, and to the owners of the corporation—the shareholders. On the new model, however, they are not responsible to the shareholders for maximum profits. The desire of the speculator may be for maximum profits, or even more accurately for a significant increase in the price of the corporation's stock. But the board is not responsible to speculators for stock price increases or for anything else. To those shareholders who know or care about the corporation and who are not merely speculators, the board is responsible for overseeing management and evaluating its effectiveness in achieving corporate goals. Obviously, one of these is profit. But profit is not the only goal, and investors have no right to think that it is.

Management is responsible to the board. But it can obviously be held responsible only if management is different from the board.

The second organizational demand is, therefore:

2. Require that not more than half the board come from management ranks.

Workers may be represented on the board, chief stockholders might be, there might be some members from management, and some members from outside the corporation entirely. Management can clearly be held responsible only if management is different from the board. Executive compensation and perquisites, such as stock options, are notorious areas for conflict of interest. Management can hardly be expected to evaluate its own performance objectively or to compensate itself without prejudice. The pattern of outside board members or of a majority of outside board members has been established in Europe[7] and is a pattern that is being followed by some corporations here. It is a pattern that fits the new model.

Responsibility on lower levels is assignable and the assumption of responsibility should be not only required but taught.

Responsibility without accountability, however, is empty. It is here that responsibility and responsiveness join together. Accountability requires not only a response to those to whom one is responsible, but it also requires access to information and the giving of reasons for decisions. One of the ways to preserve one's power is to monopolize information so that others have to assume the decisions taken were proper in the light of all the information. Accountability under such conditions is impossible. Thus accountability requires disclosure. There are certainly legitimate areas for corporate privacy to hold sway, such as having to do with the strategy of corporate research or the details of trade secrets, and those should be protected by law. But such areas are probably fewer than most corporations are willing to admit. If people are to be held responsible for their decisions, those affected by a decision have a right to know at least something of the rationale for it. The supply of information and of reasons for a decision do not mean that the person who made the decision had no right to make it. But knowing that one may be called on to explain or defend decisions helps them from being made arbitrarily. Hence,

> 3. Determine how much disclosure is appropriate at each level and to whom. The determination should be made not unilaterally, but through reasoned discourse, in conjunction with those seeking information and those to whom one is rightly accountable.

Accountability as I see it developing under the new model is not simply hierarchical. Management should be accountable to the workers as well as to the board, and the workers should be accountable to each other as well as to management. The board should be accountable to the stockholders and to the public, but it should also be accountable to management, providing reasons for the decisions it takes. The rationale, if it needs spelling out again, is that the new model sees the corporation as composed of people and not simply of functions or positions on an organizational chart.

Accountability as I have described it here is the heart of both the demand for responsibility and for responsiveness, and it requires significant organizational modifications. Hence,

4. Establish channels and procedures for accountability up, down, and laterally.

There must be ways by which workers can make known their demands and concerns without fear or prejudice and receive explanations for decisions that affect them. Clearly, to speak of channels and procedures goes well beyond the suggestion box. The creation of a position of ombudsman has been tried in some corporations; special departments might be established in other corporations. Accountability to the public, moreover, is the chance for corporations to tell their side of the story, to indicate what social demands they have considered, why they have made the decisions they have made. The American public does not expect corporations to act from moral motives. It does expect them not to violate basic moral rules and to consider the social ramifications of their actions. If many corporations do, they fail to report this adequately or convincingly. Hence,

5. Develop input lines whereby employees, consumers, stockholders, and the public can make known their concern, demands, and perceptions of a corporation's legitimate responsibilities.

6. Develop a mechanism (large corporations shold develop a department) for seriously considering and weighing the various demands, for anticipating them, and for proposing appropriate action.

7. Develop techniques for disseminating to those interested the basis for decisions affecting the general good.

The mechanism for weighing various demands is crucial. Many of the demands, I have already indicated, cannot be made in cost accounting terms. Where they cannot, then the arguments in defense of one set of actions and a consideration of its consequences are typically pitted against similar considerations for other sets of actions. But the only way to make the discussion fair is to have advocates for each side. For maximum effectiveness and for assurance that both sides will be heard, these advocates should come from within the corporation, even if they represent demands made by those outside the corporation. In at least some office or department it should be proper to argue not what the corporation can get away with, but what is the right thing to do. That decision will have to be weighed against cost and other factors. But unless someone is paid to argue against the company's position, unless his position and advancement are dependent on his properly and strongly presenting the case of those outside the company, and unless he has some likelihood of winning, ouside demands will not get an adequate hearing. A progressive company would demand even more, namely, that some people within it be responsible for anticipating societal demands so that the company can respond to them before outside forces have to be marshalled. The development of a group, office or department within a corporation that argues against the company's short-term interest in the light of its larger responsibilities is a major organizational modification suggested by the new organizational model.

Responsibility without sanctions, however, is empty. The demand for responsibility requires that sanctions be developed and enforced. If responsibility is personally assignable, then there is little difficulty in knowing whom to sanction. Kickbacks, foreign bribes, the use of insider information, and the like are everyday fare in the newspapers and magazines. The deceit, deception, and the dishonesty of a few tarnish the image of the many. Yet very few corporations have been willing to admit blame for their wrong-doing, fewer still take any effective measures against their managers or board members,[8] and even fewer cast stones at fellow businesspeople. It is difficult to believe that businesspeople do not know when others are acting immorally, however that is defined. Yet rarely does any businessman publicly bring charges of impropriety or immorality; and even more rarely do any self-policing mechanisms result in the imposing of severe sanctions. Hence,

> 8. Enforce responsibility with sanctions both within an organization and, to the extent possible within the bounds set by anti-trust legislation, across an industry. The price for executive irresponsibility or immorality should be as severe as for lower employees.

The third facet of responsibility involves conflict resolution in those instances in which a company policy comes into conflict with the moral views or norms of the individual. There should be some mechanism whereby this discrepancy can be dealt with without endangering the position of the person raising the objection. There are a number of famous cases of whistleblowing in industry,[9] and the whistleblowers have in most cases fared poorly for their moral stance. The reason seems to be built into the traditional view of an organization embraced by most corporations. Employees are to do what they are told and are not to ask questions or cause trouble. If they see defects or immoral practices, they may in some cases report them to those above them; but once they have done so they have discharged their corporate responsibility; and if they do not report them, they are not morally responsible to do so. This view is doubly at fault. To the extent that employees are not expected to let their moral concerns enter into what they do, the organization in fact depersonalizes them. To the extent that a corporation has no mechanism for considering the moral implications of its actions as seen by its employees, it wishes in effect not to consider the moral dimension. Such a corporation can hardly be surprised when it is faulted for being nonmoral or immoral. Hence,

> 9. Preclude the necessity of whistleblowing by providing procedures, mechanisms, and channels whereby any member of the organization can file moral concerns of the type that lead to whistleblowing and can get a fair hearing and possible action without fear or negative consequences.

To give this teeth:

> 10. Hold some highly placed official in the corporation responsible for paying sufficient attention to legitimate claims about product safety and the like.

To have procedures for handling conflicts of corporate policy and employee morality does not mean that anything that anyone claims to be immoral is immoral. But there should be organizational procedures so that such charges get a full and fair hearing from those who will ultimately be responsible without threat of negative consequences to those raising the issues. As opposed to the traditional model, people at every level should be held responsible for seeing and reporting such issues and penalized for failure to do so. If industry were responsive to legitimate complaints by those who see product dangers, employees would not have to go public to get corporate action.

Conclusion

I have suggested that a new model of the corporation is emerging and that it carries with it certain organizational changes or imperatives. The democratic process does not lead us to expect that we will always get what we demand; but it does lead us to expect that our demands will be rationally considered, and, if justified, met. Democratic humanism has moved from the political realm to the marketplace. The new view of the corporation reflects this move. The organizational changes I have suggested provide the mechanism for an adequate response. For if a corporation is to face conflicting social demands squarely, it must be able to learn clearly what they are and do so early enough that it can take the most effective action. It must develop a mechanism for weighing conflicting demands. And it must have the means not only for implementing its decisions, but also for explaining and defending them to those whose demands could not be met.

The changes taken as a whole fall far short of socialism and workers' self-management. I do not believe there is a mandate yet either to go entirely in that direction or to go that far. Because of this, some may accuse me of defending the status quo. Others, enamored of the traditional model and unwilling to give up any of the traditional privileges and autocracy of management, will accuse me of calling for an end to free enterprise and capitalism.

The changes I have outlined are not all original and most have been tried somewhere or other in some form or other. But they seem to me to form a whole, to come from a conception of what a corporation should be, and to be in line with what is being demanded of business. Making the changes will in many cases not be easy. But whether the corporation changes in this way or in some other way, change it must if it is to survive. The longer it waits, the less likely it is to emerge in anything like its present form.

Acknowledgment

Portions of this paper appeared in "Responding to the Mandate for Social Responsibility," *Guidelines for Business When Societal Demands Conflict* (Washington, DC: Council for Better Business Bureaus, 1978) pp. 60–80.

Notes and References

[1]"Moral Issues in Business," *Ethics, Free Enterprise, and Public Policy*, R. De George and J. Pichler, eds. (New York: Oxford, 1978).

[2]For a good survey of what some corporations have done, see John L. Paluszek, *Will the Corporation Survive?* (Reston, Va.: Reston Publishing Company Inc., 1977).

[3]Leonard Silk and David Vogel, *Ethics and Profits* (New York: Simon and Schuster, 1976).

[4]See, for example, Amitai Etzioni, *Modern Organizations* (Englewood Cliffs, NJ: Prentice-Hall, 1964); and Herbert A. Simon, *Administrative Behavior*, 2nd ed. (New York: Free Press, 1965).

[5]This has been developed by John Ladd, "Morality and the Ideal of Rationality in Formal Organizations," *The Monist*, 54 (1970), 488–516.

[6]There is a growing literature on the social audit. See David F. Linowes, "The Corporate Sociate Audit," in *Social Responsibility and Accountability*, Jules Backman, ed. (New York: NYU Press, 1975); *Corporate Social Accounting*, Meinolf Dierkes and Raymond A. Bauer, eds. (New York: Praeger, 1973; and Clark C. Abt, *The Social Audit for Management* (New York: AMACOM, 1977).

[7]Paluszek, *op. cit.*, pp. 67–68.

[8]The 1974 DC-10 case is a classic. See Paul Eddy, Elaine Potter, and Bruce Page, *Destination Disaster* (New York: Quadrangle/The New York Times Book Co., 1976), especially pp. 283–284.

[9]See *Whistle Blowing*, Ralph Nader, Peter J. Petkas, and Kate Blackwell, eds. (New York: Grossman, 1972).

Part III

Professionals in a Corporate Setting

Professionals in a Corporate Setting

An Introduction

Michael S. Pritchard

The first two sections of this volume have treated business and professional ethics more or less independently of one another. However, for an increasing number of professionals there can be no firm line dividing the ethical problems of business and the professions, simply because more and more professionals are employed by corporations. This is obvious enough when we consider professional managers, accountants, data systems analysts, and marketing specialists. But even traditionally self-employed professionals such as engineers (the vast majority of whom now work within corporations, and many of whom move into managerial positions), lawyers, and medical professionals are increasingly in the employ of large businesses.

This intimate relationship between professionals and the business world has given rise to a number of serious problems, problems that pivot around actual and potential conflicts between the aims and goals of business and the aims and goals of the various professions. For example, many engineers are members of professional engineering societies that subscribe to codes of ethics insisting that engineers hold paramount the duty to protect, if not promote, public safety and welfare. At the same time, an engineer may be employed by a corporation whose leaders are committed to Milton Friedman's view that the only social responsibility of business is to maximize profits for stockholders. What should happen when an engineer is convinced that a product poses an unreasonable risk to the health or safety of the public? If putting the product on the market would be profitable and not illegal, a decision to market the product might well be made. If we cannot assume that the law prohibits the marketing of every product that poses a serious health or safety hazard to the public, the engineer is faced with a serious ethical question. Should he or she be prepared to "blow the whistle" if management is not responsive to the engineering judgment that the product poses such a hazard? Or should the engineer, perhaps from fear of being fired or from loyalty to the employer, keep silent and simply do what he or she is assigned to do?

William May's essay, "Moral Leadership in the Corporate Setting," captures the essence of this kind of problem facing professionals: "The professional and the professional manager *belong to* the institution, even when in positions of top management and control. The institution is hardly subordinate to the professional as superordinate." Furthermore, May argues, most professionals are not well-prepared to face the moral complexities of life in a corporate setting. They are presented with outmoded codes of ethics, if any at all. And they are sometimes subjected to pressure from their employers to violate the codes. Furthermore, most discussions of professional ethics betray a latent individualism that does not address itself to the institutional structures within which the professional operates. This individualism is also expressed in the codes of ethics, which typically prescribe principles more suited for an independent consultant than a corporate employee.

May's essay is an attempt to redirect our attention to the institutional setting within which professionals work. After outlining the basic features of this setting, May argues that there is a need for transformational rather than simply transactional leaders in business. Such leaders would be concerned to anticipate moral problems rather than simply react to quandaries and crises after they arise. This transformational approach, May says, is the functional equivalent for business leaders of preventive medicine or legal advising.

Thomas Donaldson's "Accountability and the Bureaucratization of Corporations" addresses itself to some of the barriers increased bureaucratization poses for the accountability of professionals in corporations. Donaldson identifies four major problems. First, there is a correlation between increased bureaucratization and increasingly impersonal rules. Accountability tends to get submerged in rules, inviting rejoinders such as "I was only following the rules." Second, with greater bureaucratization there is a tendency toward more centralized decision-making, with a corresponding distance separating centers of power from major activities of the corporation outside those centers of power. This increases the chances that those with the most power lack adequate knowledge of important consequences of activities away from the centers of power. Third, greater bureaucratization, through increased specialization, results in the isolation of various strata in the corporate hierarchy. Fourth, Donaldson argues that some of the ''technocratic'' professions often lack the spirit of altruism or service that characterizes the more traditional professions such as law and medicine.

Although Donaldson does not claim to have a full remedy for the problems he identifies, he does suggest that greater democratization of the corporate workplace might help. He suggests a participatory model of accountability. An important feature of this model is the recognition that employees should have the right to ''blow the whistle'' concerning dangerous products or unsafe working conditions—without having to suffer penalties or job loss. The topic of whistleblowing is pursued at some length in several of the essays that follow.

John Kultgen's essay is an in-depth analysis of codes of ethics, particularly engineering codes of ethics. Kultgen argues that it is important to subject codes of ethics to close scrutiny because of the impact of the professions on our vital interests and because of the increasing professionalization of occupations in our society. It is characteristic of emerging professions that are striving for recognition to develop codes of ethics. Kultgen's concern is to examine codes to determine to what extent it is desirable to have codes and to determine what content it might be desirable for codes to have. He is careful to point out that codes are interpreted differently by different audiences, and the designers of codes may take this into consideration when they frame them. Thus, the vagueness and indeterminacy of engineering codes allows them to be interpreted differently by, for example, the uninformed public that assumes there is a single engineering profession, an elite group of consulting and managing engineers, and engineers occupying subordinate positions in large organizations. Kultgen emphasizes the different kinds of functions that can be served by codes. For example, they might serve social, regulative, or ideological functions. He proposes that codes of ethics be evaluated philosophically from a rule-utilitarian standpoint. Thus a code is to be judged according to whether it promotes a social/institutional framework that tends to maximize human happiness.

Mike Martin's ''Professional Autonomy and Employer's Authority'' examines the apparent conflict between being a loyal employee and being a morally sensitive professional who is concerned with the public good. Martin rejects the view that loyalty to an employer and recognition of the authority of an employer require accepting directives without subjecting them to critical review. And he rejects the view that the professional's moral obligation to the public requires that professional activities simply conform to his or her independent calculation of their consequences for the public good. Although such calculations should be made, it must be acknowledged that professional judgments may differ. And a further judgment must be made as to whose judgment should prevail. Martin argues that an analogy can be drawn between being a loyal citizen and being a loyal employee. Both require respect for legitimate authority and the public good. But neither allow blind obedience.

Martin's essay emphasizes the responsibilities of professionals. The remaining essays shift attention to rights and liberties that should be accorded professionals, particularly if they are expected to fulfill certain responsibilities. Robert Ladenson argues that there should be freedom of expression in the workplace. Many have complained that in the corporate workplace there are tremendous pressures to conform and refrain from criticism. Although whistleblowers are provided some protections, employers in private corporations have considerable discretion in hiring and firing employees.

Ladenson deliberately restricts his attention to the moral case that might be made for freedom of expression. This, he claims, is because the moral case needs to be argued before arguments are made for acknowledging a legal right to freedom of expression in the workplace. Ladenson explores two approaches. The first argues for the need for employees to be "volunteer public guardians" whose responsibility is to benefit society by providing protection from harm. The effectiveness of this argument depends on there being a causal connection between freedom of expression and the provision of these benefits. The more marginal the benefits, the less strong is the argument for freedom of expression—and the less protection will be provided the whistleblower.

The second argument, which is the one advocated by Ladenson, appeals to John Stuart Mill's *On Liberty*. Freedom of expression can be understood to be necessary for the development of individuality. As Ladenson puts it, this freedom is a part of the "distinctive endowment of a human being." Any reasonable conception of the good for society, Ladenson argues, would acknowledge that a primary function of social arrangements should be to aid people in cultivating their individuality as much as possible. Unlike the first argument, this second argument does not place less value on freedom of expression if there is a greater likelihood that a whistleblower's statements are mistaken.

Gene James argues for legal protection for whistleblowers. Although loyalty to employers is the most frequently given reason for not blowing the whistle, James thinks that the most basic reason employees refrain from whistleblowing is self-interest. There is a fear of losing one's job or being subjected to sanctions. There is also a fear that there will be a violation of the law because of legal duties of confidentiality and loyalty. Finally, James says, there is relatively little legal protection available for employees who do blow the whistle.

In general, James' essay offers a realistic appraisal of the prospects for whistleblowers, given our present laws and the status of professional societies. Although not advocating irresponsible whistleblowing, James does believe that some whistleblowing is necessary if professionals are to fulfill the kinds of responsibilities appropriate to their professions. So, he advocates increasing legal protections for whistleblowers, and he urges professional societies and unions to provide support for those who lose their jobs as a result of conscientiously informing others of unreasonable health and safety hazards posed by the corporation for which they work.

The concluding essay, Albert Flores' "On the Rights of Professionals," argues that there are certain rights possessed by professionals as a consequence of their professional status. Flores points out that most codes of ethics (as well as most discussions of professional ethics) emphasize the duties and responsibilities of professionals, while failing to mention the rights that professionals must be accorded if they are to fulfill their responsibilities. Flores argues for these rights from within a natural rights framework. This framework, Flores maintains, assumes an intimate connection between having rights and having a valued status. Rights are invoked to protect certain activities or interests that are essential to this status. Professionals, as professionals, are accorded a certain status in society. This status is related both to their expertise and to the responsibility to use that expertise in ways beneficial to society. However, Flores argues, if this status is to be respected, it is necessary to acknowledge that professionals have certain rights by virtue of their professional status. If these rights (such as the right not to have their professional skills misused or abused by their employers) are not respected, professionals will not be able to fulfill their responsibilities.

The essays in this concluding section reveal many of the moral complexities of professional life in a corporate setting. They make clear the need to address the problems of professional ethics within a business environment. Even if it is conceded that the goals of the professions are different from the goals of business, it cannot be ignored that much of the life of the professions is inseparable from the business world. This inseparability gives rise to serious problems, and at times it may threaten to compromise the goals of the professions. On the other hand, as some of our authors have maintained, the presence of ever-increasing numbers of professionals within the corporate setting also provides opportunities to encourage corporations to embrace the social responsibilities advocated by the professions.

Moral Leadership in the Corporate Setting

William F. May

Introduction

This essay will attempt to deal with the moral problems faced by the professional and the professional manager at work in the corporate setting. More and more professionals work today for large-scale institutions, public and private, under whose authority they generate professional services. These institutions fall into two classes: those that still serve the fundamental purposes of a profession (schools, universities, hospitals, law firms, newspapers, consulting firms), and those whose purposes treat professional goals as instrumental (most notably, the government and the business corporation). Not only do more and more professionals work for large-scale organizations (the institutionalization of professionals), but, increasingly, professionals head them (the professionalization of institutions). Specifically, the task of management in these organizations has been professionalized. Corporations and service bureaucracies recruit most of their leadership either from the traditional professions or from a growing company of administrators who acquire formal training in management at the universities. Although top jobs in management do not go only to those with a professional school education, mere amateur graduates of liberal arts colleges know, in fact, that by other routes few arrive at the executive suite.

The increasing concentration of professionals in huge and often commercial organizations has profoundly affected professional practice. It permits professionals to specialize to a degree not possible before; it affects greatly their specialties; it generally encourages a higher standard of technical performance than the solo practitioner can achieve. Identity with a large-scale organization also endows the professional's work with a kind of public significance that it does not acquire in more intimate settings; and it has magnified the professional's power over the society at large.

Yet troubles beleaguer professions, along with the institutions they lead. Many of the professions have overproduced practitioners[1]; laypeople have challenged professional authority; public scandal has shaken assumptions about professional rectitude in public life; their gu ilds have often shown an inability to keep their own houses in order; an d, increasingly, people hold professionals responsible for the failures and defects of those institutions that they serve, advise, and often control.

Professionals do not find themselves particularly well-equipped to face the moral complexities of life in a corporate setting. Their professional associations provide them either with outmoded codes that, in the words of a lawyer, were made for "downstate Illinois in the 1860s,"[2] or with no codes (in the case of most managers), or with professional societies that (in the instances of engineering and nursing) offer little support to beleaguered professionals in moral conflict with employers who would pressure them to violate their professional standards. Some professionals, of course, resort to formal academic ethics for help in facing the discrete quandaries they face, but, on the whole, academic ethicists offer little discussion and criticism of the institutional structures within which the professional operates. This latent individualism vitiates ethical analysis. Systems, institutions, and structures shape in advance the horizon against which the professional operates and the problems that surface as cases. Unless the ethicist attends to that institutional horizon, he or she systematically ignores a fair portion of the task of moral criticism.

The Marks of the Professional and the Professional Manager

Abraham Flexner, whose influential Carnegie report on medical education (1910) gave him near judicial authority in the determination of professional status, settled on three basic components of professional identity: intellectual, moral, and communal/institutional.[3] Intellectually, the professional draws on a body of complex, theoretical, and esoteric knowledge that a university education offers. This knowledge itself grows, changes, and amplifies through the research of the profession, which, once again, the professional usually acquires in the university. Morally, the professional derives great power from this knowledge, but, unlike the wizard of old, does not wield it for purposes of self-display. Rather, he or she uses this knowledge-based power to serve human need and applies it to solve concrete problems. Although professionals take pay for their work, altruism rather than vulgar self-display or crass commercialism should motivate them. Thus knowledge links power to philanthropy. Finally, professionals organize themselves into distinctive guilds charged with maintaining professional standards. As befits the moral mark of a profession, these guilds should differ from trade and commercial associations by dedicating themselves to moral self-improvement rather than to self-promotion. As befits the intellectual mark of a profession, professionals should organize collegially rather than hierarchically. Grasping theoretical principles, each professional should command his or her own direct access to the knowledge base from which professional authority derives. So goes the professional ideal.

Not surprisingly, Abraham Flexner denied professional status to those who managed business institutions. In Flexner's day, managers hardly drew on a tradition of theoretical knowledge available through a university education; they pursued cash rather than the ideal of service; and whatever guild associations they established encouraged self-promotion rather than self-regulation. Measuring them against his checklist of traits, Flexner felt compelled to exclude business people and managers from the professional class.

But both managers and professionals have blurred Flexner's clear-cut line between the commercial and the professionals in our time. Professionals have learned how to convert their careers and guild organizations into instruments of commercial advantage and tribal self-protection. Managers, meanwhile, increasingly aspire to meet Flexner's intellectual and moral standards. Although they have not established guilds with independent principles and enforcement powers, they increasingly emphasize the importance of a professional school education and they describe their enterprise as a "public service." Only a cynic could doubt that the ideal of service will become as weighty a component in corporate policy as in advertising appeals. Still, professionals and managers today seem to differ little from one another in intellectual and moral characteristics.

Others have sought to distinguish professionals and managers along somewhat different lines—functionally rather than intellectually and morally. Both the professional and the manager function in asymmetrical social relationships; both exercise some power and authority as superiors over subordinates: the professional over clients, and the manager over employees. However, the professional superordinate acts exclusively on behalf of the client's or patient's welfare; whereas the managerial superordinate exercises authority over the subordinate on behalf of the institution that they both serve. Clearly the manager must consider the welfare of workers under his or her charge, but the relationship does not exist—at least not primarily—for that purpose. Both serve the well-being of the institution from which their own welfare mediately derives.

Walter Metzger, the Columbia historian, presses this functionalist view in his essay "What is a Professional?",[4] but Metzger wrongly restricts the term "professional" to those who serve clients. His functionalist definition overlooks those important professions that invoke the ideal of service and that rely on the intellectual competence of their practitioners, but that, strictly speaking, serve no clients—the civil, military, and foreign services. These traditional 19th century professions, along with the somewhat more complicated profession of the ministry, chiefly serve institutions that, in their own right and in turn, theoretically contribute to the larger social good.

Still, Metzger's functionalist distinction highlights the special problem faced by all professionals at work in an institutional setting. The professional and the professional manager *belong to* the institution, even when in positions of top management control. The institution supersedes the professional as superordinate. The modern manager subordinates himself or herself to the purposes of the institution and finds it difficult (almost as difficult as the civil servant and the military leader) to maintain independence, whether psychological, moral, or professional from the institution and its imperatives. The contrast in the sheer length between case studies in medicine and business symbolizes the difference between the traditional professional and the manager. Cases in medical ethics are remarkably terse. Doctors usually leave their patients dead by page two. In contrast, cases in management ethics plod along interminably, some for fifty to seventy-five pages. This difference in length symbolizes an important difference in the moral problem each faces. The physician may treat thousands of patients. Sheer numbers allow for, and indeed, demand some distance from each. The physician develops calloused detachment and faces some of its attendant moral problems. But the manager must live, breathe, worry, anticipate, and aspire largely within the confines of the single corporation for which he or she works. It envelops the manager. It becomes his or her *de facto* world. This psychological envelopment does not confer upon managers the right to abandon either universal moral constraints or the special constraints that managers, as a professional class, may see fit to accept for themselves. But it does mean that the very structure of the institution and its purposes must impose whatever moral constraints operate. Management must accept as an essential responsibility this task of moral construction. Morality cannot depend entirely on the fitful efforts of individual and heroic persons. Inevitably, management ethics must begin with the institution, its mission, and its structure.

The Marks of the Corporation

It would take a very large organization to house all the literature generated on the subject of the large-scale organization. Nevertheless, the literature describes a few features that provide a beginning for ethical analysis. Large organizations usually define themselves by a single stated primary mission (health, education, or the sale of products for a profit). Furthermore, they fall into hierarchical and internally differentiated structures; this internal differentiation means that procedures inevitably regularize from department to department. Finally, large-scale organizations emphasize the office rather than the person. They specify roles and functions and classify cases rather than highlight the man or woman filling the role or presenting the case.

Large-scale organizations can deviate from these characteristics. They can develop complex primary ends (the university teaching hospital simultaneously heals, teaches, and sponsors medical research; the conglomerate spans a number of businesses); but this complexity usually serves a single, inclusive primary aim (health or economic performance at a profit). Further, the pyramidal structure of the corporation can be flat or steep; it can be decentralized or federal[5]; but even the decentralized large-scale enterprise does not altogether dispense with hierarchy and routine. Finally, it may attempt to personalize service and humanize the work environment, but cannot, because it serves categories and masses, dispense with impersonal standards of treatment and behaviors. Some deviation exists; the marks of the corporation are hardly indelible, but they persist enough to justify directing moral reflection to them.

Primary Institutional Purpose

The first mark of a large-scale organization is its tendency to orient itself to a single purpose (health, education, economic performance). This tendency to specialization has attracted both fierce critics and passionate defenders. At one end of the spectrum, critics from the left in the late sixties protested against major organizations that kept to their specialized tasks, but failed to respond to overriding moral challenges and political crises that the radicals felt should test an institution's right to survive. At the other end of the spectrum, conservatives, but not only conservatives, defended the view that a society functions best if each institution within it pursues its own special mission. The term "conservative" here loosely covers two very different visions of the social order. Classical conservatives construct an organic image for the body politic and envisage the great institutions of the society working together in compact harmony for the common good (Japan incorporated). Contemporary American conservatives, led by Milton Friedman, argue for a minimalist state and believe that a multiplicity of goods will best develop through the allocative wizardry of the free market. Far from subordinating self-interest to some notion of the common good, business institutions, in this libertarian view, can make no greater contribution to the society than to pursue, without distraction, the maximization of profits.[6] Otherwise, corporations exceed their authority (derived from stockholders), their competence (economic decision-making rather than social engineering), and their power (as they paternalistically impose on others their notions of social and cultural good).

Neither the New Left nor the American libertarians (I exclude from this discussion the classical conservatives) offers an altogether satisfactory resolution to the problem of institutional purpose. The ideal of multipurposed or polytelic institutions overlooks the very prosaic advantages of specialization: efficiency and productivity. For all its talk about community, the New Left undercuts the human impulse toward reciprocity. A latent hostility to reason and community lurks behind the New Left's reluctance to specialize and its refusal to accept what specialization implies, a dependency upon the dedicated efforts of others. Mutual dependency in a differentiated society requires that institutions undertake a primary mission, a mission, moreover, that it must sustain across an extended period of time. The Left is impatient with extended time both for persons and institutions, and reduces everything to the moment. Anarchy results if institutions universalize themselves and try to do everything. The educational institution must educate well; it cannot, in and of itself, respond to any and all moral and political challenges served up to it before lunch on any given day.

But Friedman's monomania about the maximization of profits does not work as a definition of corporate purpose. Life in a mixed economy requires performance at a profit; but the maximization of profits at the expense of all else converts a requirement into an obsession. Friedman, to be sure, accepted the side constraints of ethical custom and the law in his essay on the social responsibility of business. But he holds to a highly individualistic conception of ethical custom (no lying, theft, or murder) and a minimalist understanding of the law. He hardly means by ethics, social ethics, and he hardly wants from the law much more than those protections that permit one to push profits to the hilt.

Even at a crude self-regarding level, it behooves the corporate world to accept social responsibilities beyond maximizing profits. Large institutions cannot afford to neglect their secondary responsibilities to their workers and to the neighborhood and society about them. Corporations, universities, and hospitals eventually suffer if they wield the power of giants in their field, but pretend to social and political impotence and let the cities around them fall to pieces. Ironically, the sheer size of these institutions tends to obscure these obligations. Their grandness of scale bestows a kind of public meaning on the lives of their workers, drawing them out of a cramped privacy. This grandeur, however, creates two problems. It obscures for participants the fact that the institution's imperatives do not exhaust the meaning of public responsibility, that the institution and its goals must take their place within a still larger framework of common good. Further, sheer size bestows on the organization the sheen of immortality. Its majesty makes appeals to longe-range peril seem, at best, remote. Short-run pressures to maximize profits for stockholders will tend to overwhelm appeals to long-term peril unless other moral considerations accompany the latter appeals. Whether the corporation will ever suffer from its sins of omission and commission remains historically problematical. Why not maximize profits now and let the future take care of itself? Money, after all, is portable—from Northern cities to the South, and from Southern cities to overseas investments. But the prior moral question must ask whether the corporation must in the first instance undertake duties to more than its stockholders. If it need not, then all efforts to expand the notion of corporate responsibility beyond the goal of maximizing profits will seem, at best, peripheral and problematical, and at worst, illegitimate, unauthorized and immoral. The ideal of corporate responsibility will seem but to squander stockholders' resources for the protection and benefit of those who have no rightful claim to these holdings.

Some critics have sought to counter Friedman's first argument from legitimacy by distinguishing between stockholders and stakeholders. Stockholders are important stakeholders in a company, but by no means the only ones. Workers, customers, neighbors, and the public at large have in varying ways a stake in its performance, sometimes indeed a larger stake than stockholders who may dart in and out of their investments more readily than workers and neighbors can disengage themselves from a company and its fortunes.

This argument, however, rests on little more than the consonance of the two words—"stock" and "stake"—unless one can show why obligation should extend beyond stockholders in an enterprise. The notion of a covenant at once expands corporate obligation beyond the limits of a commercial contract (the stockholder's purchase) and takes corporate responsibility a notch higher than token expressions of corporate philanthropy (contributions to the local Community Chest drive). Very briefly: a convenantal ethic concentrates on those obligations that arise in responsible and reciprocal relationships between several parties. It acknowledges a two-way process of giving and receiving that governs extended exchanges between parties, from which permanent agreements arise that shape the future. This growing responsibility between parties takes practical form in the notion of a stakeholder. Much has happened between us that gives me a stake in what you do, and you, in what I do. We are covenanted together.

In contrast, the ideal of philanthropy presupposes a one-way street from giver to receiver. The philanthropist pretends to give only; others receive only. When the philanthropist loses interest, recipients must move on. Thus corporate responsibility, when reduced to philanthropy, trivializes corporate ethics. The corporation obscures the depth of its obligation when it adopts this view. It does not give as pure benefactor alone. It has accepted much from the community—not only the investments of the stockholders, but also the labor of workers, the ambiance and services of the community in which it works, the privilege of incorporation by the state, and all the protection bestowed upon persons under the due process clause of the constitution. It owes a great debt to the society.

But not all this indebtedness can be summed up in commercial terms. The corporation, of course, pays its workers and its taxes, but these transactions build up a life between people and institutions, rather than terminate a connection. Thus the word "covenant" rather than "contract" best defines the moral obligations of the corporation. Formally considered, "contract" and "covenant," like first cousins, share some family features: they both include an agreement between parties and an exchange; they look to reciprocal future action. But, in spirit, contract and covenant differ. Contracts govern only buying and selling. Covenants include further ingredients of giving and receiving. Contracts govern economies; covenants permeate lives. Contracts include provisions for expedient discharge; covenants nourish rather than limit and terminate relationships. Contracts end in files, covenants become part of one's history and shape, in unexpected ways, self-perception and perhaps even destiny. The decision-making of a corporation is massively contractual. But that decision-making rests upon an unacknowledged covenantal base that charters its life, grants it protection, and endows its enterprises with a public significance and responsibility.

Friedman's second argument—that corporations are incompetent to make judgments about matters other than profit-making—underestimates both their competence and their responsibility for competence. In some areas of corporate responsibility (for example, product safety and negative effects of manufacturing processes on workers and neighbors) management usually has accumulated more knowledge and competence than most others in the society. In other areas, the corporation may not yet judge competently. But moral responsibility in life is hardly limited to areas in which persons and institutions have an advance competence. Competent lovers may find themselves—willy-nilly—somewhat nervously in the role of inexperienced, not-yet-competent parents, but must morally and practically acquire that competence. Economists have conveniently excluded many issues from the manager's responsibility for competence by writing them off as "externalities." But the externalities of water and air pollution, traffic congestion, neighborhood crime, unemployment, and the quality of education and culture in a host city obtrude on the moral and political agenda of citizens and their institutions and require them to develop some measure of wisdom and even expertise.

Finally, critics charge advocates of corporate responsibility with paternalism—that is, with a well-meaning but nonetheless oppressive interference in the affairs of others. The very word "paternalism," of course, stirs a subtle mix of negative memories, including Puritan forebearers, Victorian company towns, the horror of total institutions, and resentment against a pampered but oppressive childhood. Such critics believe that the corporation, precisely because it exercises great economic power, should forswear further social, political, and cultural interventions.

Undoubtedly, paternalism is an evil. It diminishes human autonomy and freedom, in the name of benevolence towards others. Modern industrial cities, however, suffer less from the overbearing presence of benevolence than from its absence. The unbridled self-interest of the powerful threatens human autonomy and freedom today far more than misguided good works. Interventions, moreover, need not inevitably diminish the freedom of others; they may also expand opportunity. Freedom, when responsibly exercised, does not, like a commodity of fixed bulk, diminish for some when exercised by others. The civil rights of citizens, their economic opportunities, and their cultural resources may expand and flourish to the degree that intermediate institutions in the society open out beyond their immediate self-interest toward the common good.

Even if one accepts these arguments and broadens corporate responsibility to include, in addition to stockholders, workers, neighbors, consumers, and the public interest, the question of how to order these responsibilities remains unresolved. Some have argued a strict ordinal ranking: profit, primary; care of workforce, secondary; and other social goals, tertiary. In this view, one cannot move on to secondary and tertiary goals without satisfying the primary goals of economic survival and growth.[7] In the same spirit, others have argued that tertiary goals should only help the manager choose between alternative, equally profitable, courses of action.[8] If profit ranks first in this ordinal sense, then the goal of social responsibility will slip to the outermost edges of the conscience. The demand for profit will tend to expand; no determinate level will ever satisfy it. Alternatively, if social impact serves only to break a tie, managers will use that tie-breaker about as often as the Vice President of the United States votes in the Senate. The ordinal system of relating the aims of business overlooks, moreover, the distinction between negative and positive duties.[9] Prohibitions against harm to employees, neighbors, consumers, and environment must function as side-constraints that set the limits upon the pursuit of profitability, rather than as after-thoughts that come into play only after profitability has done its uttermost. Profitability can reinforce responses to prohibitions against harm. When a rogue competitor secures an unfair advantage through socially injurious action or when negative social impacts destroy a business opportunity, managers should advocate such regulation as will deprive the unscrupulous competitor of unfair advantages. An ordinal ranking of the aims of business thus overlooks the complicated, overlapping relations between primary, secondary, and tertiary goals.

Other interpreters order these various ends serially, arguing that profit-maximizers prevailed until the 1920s, trustee managers, through the 20s and the 30s, while quality-of-lifers have acquired increasing power today. This rather benign view trusts too much to the inevitability of history. These commitments, rather than succeeding each other, accompany and often conflict in the corporate conscience.[10]

The Hierarchical vs the Collegial

The second mark of the corporation, its hierarchical design, relies upon patterns of super- and subordination. This feature places it at odds with the natural social structure among traditional professionals, which is robustly collegial. Professionals historically have accepted patterns of super- and subordination among themselves only as temporary phases of training and education. The apprentice is subordinate to the master only because he or she has not yet attained full professional skill. In achieving professional status, the apprentice acquires an independent relationship to sources of knowledge and accumulates sufficient experience to apply this knowledge to specific cases. Thus, independence marks the professional. The principle of collegiality expresses this independence within a community of professionals; colleagues act in concert with one another chiefly by persuasion rather than command.

A hierarchically ordered corporation sometimes indulges in leadership by persuasion (teaching), but finally rests upon command and upon sanctions vested in that command. The professional thus faces, in principle at least, serious potential conflicts between the imperatives of the organization and those derived from the aims and purposes of the professions. As members of a bureaucracy, they obey orders; as professionals, sometimes they must, awkwardly enough, disobey orders. The conflict goes deeper than the occasional overt crisis when the professional must "blow the whistle": the collegial and the bureaucratic types of social organization establish somewhat conflicting notions of duty and modes of social identity that one and the same person may suffer unresolved. Professional guilds vary greatly in their willingness and power to back members in valid challenges against the institutions for which they work. The American Association of University Professors has established investigatory and legal defense funds to protect its members and is willing to apply profession-wide sanctions against offending institutions. Other professions (such as engineering and nursing) have proved timorous by comparison, leaving the individual professional to brave a conflict alone—and thus rarely.

Business managers (and professionals wholly defined by institutional service—the military, civil, and foreign services) lack the leverage of professional guilds. They must, as indicated earlier, build moral responsibility into the very structure of the institution. But the hierarchical pattern of authority—quite apart from the pressures of profit-making—does not encourage the cultivation of public conscience. Most corporate managers rise from the lower ranks of the organization or others like it. Although, as leaders of the corporation, they exercise enormous quasipublic powers, their time in the lower ranks has not adequately prepared them for this responsibility. Admittedly, their advancement in the corporation depends not only upon their technical competence in their specialty, but also upon the development of general social skills. They must get along reasonably well with colleagues, superiors, and subordinates. These social skills constitute a kind of political art, the ability to act in concert with others, but they have learned an art of politics shorn of its object: the common good. Only those at the apex of organization have the right to act upon behalf of the organization or the still-wider public good. All members within the organization must submit to the decisions of top management without direct and regular voice—either in policy-making or in the selection of those who make policy.

Put another way, workers and most lower to middle level managers belong to the organization, but they enjoy no citizenship in it. They neither make the Kissinger-like decisions that affect the well-being of the multitudes, nor do they select the Kissingers in their midst. They serve at the pleasure of management and do not possess *de jure* either the security or the responsibility of the indelible office of the citizen in a democracy. In brief, public virtue would be an anomaly in a corporate manager below the uppermost levels, and for that reason, one would not expect to see it suddenly appear in those newly appointed eagles in the enterprise who did not previously learn it in their formative years.

Even if, by accident, the head of a corporation should acquire with the office some sense of public responsibility or aspire to the mantle of statesman, the prevailing culture of the institution out of which he or she emerges often makes it difficult to restructure procedures or influence motives through the organization. Even if such a head establishes a special office for corporate responsibility, it merely advises the president discreetly. It often subsists, like a benign tumor, on the larger organism, without affecting it in structure or substance. Middle level managers will operate according to conventional canons governing performance and promotion, to which questions of public good remain, at best, marginal. Thus presidential speeches on public responsibility tend to sound like hyperbolic press releases.

This state of affairs has led some to look to external regulations as the only route to corporate responsibility; others have sought it through externally mandated reforms in corporate structure; and still others would add to these the need to affect, in more interior ways, the corporate culture. The increasing presence of professionals in the organization suggests at least one possible change in corporate culture.

Intramurally, professionals and managers will need to expand the authority of persuasion in their dealings with one another. This shift will make room for the collegial impulse in otherwise hierarchical institutions. It may help cultivate, at earlier stages in the career of managers, a sense of public responsibility. But it requires new emphasis on the professional and the professional manager as teacher.

Bureaucratic Routine vs the Professional Aspiration to Excellence

The third mark of the large-scale organization, its commitment to bureaucratic routine, can place it at odds with the professional aspiration to excellence. Because of its complex, internal differentiation of functions, an organization must regularize its procedures. To do its own work, each department must rely on the routines of neighboring departments. This routinization develops as inevitably in the business corporation as in much-maligned public bureaucracies. It frees the institution from relying excessively on the extraordinary heroism or skill of its members. Its virtue as a huge organization frees it from depending overmuch upon virtue.

Max Weber formulated this movement from poetry to prose under the rubric of the routinization of *charisma*.[11] An age of colonizers succeeds the more glittering era of pioneers; the bureaucratic follows upon the heroic. But change need not mean loss. A society can gain much when it has reduced the heroic and the extraordinary to the merely habitual. Bureaucracies provide for stability, continuity, and growth. Their reduction of labor to prescribed routine and their rationalization of procedures for decision-making give them an edge on the spontaneous ventures of the free-lance entrepreneur. Bureacracies grow apace, handling huge volumes of work and workers, clients and customers, without demanding exceptional heroism, virtue, or competence from anyone. The large-scale organization, in effect, compensates for the mediocrity of humankind. It attracts us because, as Sheldon Wolin has argued, its "achievements did not require, as religious transcendence did, a new man. Man could accomplish great things without himself being great, without developing uncommon skills or moral excellence."[12]

The routinization of behavior in all large-scale organizations, in effect, miniaturizes men and women. It permits, to be sure, specialization, and therefore promotes excellence of a technical order. But it usually opts for a somewhat narrow and quantifiable standard of excellence. This tendency to routine behavior confines boldness to a secondary arena: it encourages tactical cunning rather than a talent for fundamental design. The virtues required to ascend in the organization differ from those strengths needed at the top. Managers sense incongruence when they look back over the heads of the current leadership to the bolder days of the founders.

The professions, in part, resemble large organizations in their leveling impact on standards of performance; they use the same process of routinization. The very existence of the professions tends to regularize standards—in the selection of candidates for training, in education requirements, and in accepted norms of performance. These standards, for the most part, spare the laity from erratic and idiosyncratic treatment, whether at the hands of genius or fraud.

Ideally, however, the professions stood for something more than minimal proficiency. They aspired to excellence. The legal profession bows to this aspiration—at least formally—in its *Code of Professional Responsibility*. It distinguishes between minimal Disciplinary Rules and maximal Ethical Considerations.[13] The distinction corresponds roughly to the categories Lon Fuller staked out (in *The Morality of Law*)[14] between a so-called ethics of duty and an ethics of aspiration. The first ethic is largely minimalist, negative, and legislative in tenor—backed by sanctions. The second ethic is maximalist, positive, and esthetic in tone (as expressed in the phrase, "a beautiful piece of work")—and proposes rewards and honors. The first establishes uniform standards; the second concedes diversity in the forms of excellence. The first endures through rules; the second, largely through examples. On the whole, however, a profession tends to relax into a minimalist understanding of moral obligations.

As professionals move into large-scale organizations, the organizations tend to enforce basic standards and predictably encourage minimalism. The teaching hospital, for example, can monitor the work of its staff members; the solo practitioner can largely escape the scrutiny of colleagues. But the large-scale organization also tends to blunt the aspiration to excellence in its more daring forms. Predictable routines for handling cases work more conveniently than bold and singular responses to assignments. The modern university, for example, has so rationalized and quantified its notions of scholarly productivity as to discourage almost all risk-taking in its younger faculty. The academic profession professes a formal commitment to the truth, but it discourages risk for important truth or truths difficult to attain. Nor does the later grant of tenure change much the mindset of those who attain it. Those long inured to caution and servility in their years without tenure probably will not act boldly when they come at last into the full possession of academic freedom. One must crawl freely in order to walk freely. The ways of servility in any organization cling stubbornly in the mind. Minimal standards in a corporation sometimes provide a ground floor that attracts the uninspired to that level of performance and no more. Disciplinary sanctions can cut like a double-edged sword; they can lop off the grossly incompetent, but can also protect the mediocre in their jobs.

These difficulties that the large-scale organization faces in maintaining its drive for excellence argue for critical reflection and experimentation with alternative modes of organization. But it does not persuade one to join the nostalgic who yearn for the days of the free-lance entrepreneur. Liberals frame a pretty fiction when they assume that men and women act better in isolation than they do in society. Even the realistic Reinhold Niebuhr fell prey to liberal innocence when he titled his work, *Moral Man and Immoral Society*. Men and women need community not merely for the instrumental purpose of producing more goods and services than they can achieve by themselves, but for the moral reason of helping them act better than they could be by themselves. Others than Puritans can respect the truth in the opposite assertion: immoral man and moral society. Men and women need the support, correction, and encouragement of their fellows; and professionals are no exception.

The Official vs the Personal

The bureaucratic emphasis on the office, rather than the person holding that office, creates conflicts for the professional and the manager over duties to customers, colleagues, and oneself.

Bureaucracies emphasize the impersonal at the expense of the personal. Dealing in huge numbers and attempting to deal fairly combine to force the large organization to deal impersonally. But patients, clients, and consumers complain of impersonal treatment and the existentialists developed this complaint into a broadside against the whole of modern mass society. They bemoaned a culture that reduces subjects to objects, persons to things. The manager symbolizes this reduction, for the very term, "manager," like "manipulation," derives from the Latin word for hands; the manager "handles" others. This criticism includes an element of truth. The ideal physician not only treats the disease, but must reckon with the patient who is its host. The advertiser and the manager demean not only customers, but themselves when they manipulate them for sales alone. A bond salesman once complained that he felt uncomfortable having to view his colleagues at church and in the community as pork chops for the eating. Managing and controlling others no longer serves ways of being with them.

Still, an ordered human community has freed persons from the nasty and the brutish even when it manages, controls, and imposes, inevitably, elements of indirection and incompleteness upon contracts between human beings. Not all human relations can rise to the level of a direct, immediate "I-Thou" encounter, and the fact that they cannot need not be bemoaned. Teachers, managers, and other professionals must often accept without false dismay the incompleteness of their contacts with those over whom they exercise some control. In doing their limited jobs, they often serve persons whose fellowship they never enjoy. Indeed, if they exceed these limits, they may only succeed in making relationships forced and superficial. Personalizing relationships smacks too much of sanforizing pants. The very process industrializes intimacy. Only a false, devouring, religious romanticism strains for an "immediacy" or a "breakthrough" in every human encounter at the expense of institutional discipline and restraint.

The organizational emphasis upon the impersonal rather than the personal runs a second danger of impoverishing the self that fills the office. The careerist or the "company man" totally subordinates to the office, and turns, in the course of time, into a cipher, a spectral being, deferring satisfactions and evading pain by referring it from its deeper psychic levels to petty frustrations. The "now" generation criticized such careerism for its tunnel vision; Jungians took it to task for its one-sidedness. Careerism prefers the *animus* to the *anima,* the rational to the effective, the manipulative to the sensitive, and thereby diminishes the self. Broken marriages, alienated children, and early health problems reflect this psychic disarray.

Once again, there lurks an element of truth in this criticism, but only an element. Though institutions ought not to diminish their members, neither should they pose as the sole arena for self-realization. They cannot, and should never try to become, total institutions. Meanwhile, official roles and social masks do not necessarily impoverish the self. On the contrary, they provide the self with a social fig leaf; they permit some measure of personal life to thrive behind them.

Finally, the bureaucratic emphasis on the impersonal at the expense of the personal presents special difficulties among colleagues in America. On the one hand, Americans experience the pressure of a highly professional, bureaucratic, and competitive social structure—energized throughout by the rewards of promotion. On the other hand, they feel the pressures of a social style, predicated on friendliness as its ideal. The social structure (impersonal, increasingly hierarchical, and pressured) contradicts the social style (personal, egalitarian, and helpful). The resultant moral conflicts between loyalty to the institution and loyalty to friends can rip psyches apart.

Some societies protect their members from such conflicts by separating the public order of work from the private order of friendship. But Americans peculiarly combine these orders at the risk of corrupting them both. Caught between the demands of an impersonal social system and a social style in which friendliness counts for so much, Americans carry a heavy burden of guilt when they betray friends or when they compromise institutions. They make friends with everybody, but then find themselves making professional judgments and decisions about their friends. The intimacies of friendship inspire some measure of personal loyalty, but they also render two people more vulnerable to one another in their weaknesses. Thus we suddenly find ourselves deliberating impersonally about those in whom we have confided and who have confided in us. Uncomfortably, we turn them over to the machinery of the system—without much certainty that the system itself dispenses justice as it reviews the individual case. Social mobility increases the competitive tension. Our system lacks a ceiling above and a floor below. The rewards for backbiting can tempt; the threat of betrayal by others chills. Under these circumstances, the prevailing social style remains outwardly friendly and direct; but inwardly, wariness takes possession of the soul and corrodes the workplace with distrust.

Essays on business ethics do not often analyse these more personal moral problems faced at the workplace. The essayist normally concentrates on the more global decisions that top managers make or that middle managers confront in facing down the boss. Nevertheless, these problems demand analysis. They dominate the moral horizon for most workers and managers. They do not, to be sure, always succumb to reform. Sometimes, they do not yield much more than diagnosis. But even diagnosis offers a modest therapeutic value. It can change the atmosphere that surrounds a problem. It can remind one also that the deeper moral problems require the question not: "what are we going to do about it?", but "how does one behave toward it?" The first question presses for pragmatic resolution; the second question looks for the personal resources and virtues that go with coping. Physicians tell their patients often enough that they cannot eliminate the patient's problems, but they can help the patient to cope. Sometimes even that goal helps efforts to make very imperfect institutions work.

Three Kinds of Criticism

Critics of the large-scale organization range from radicals to regulators, from reformers to transformers. Their proposals do not necessarily exclude one another.

The Radical Critics

The social and political program of the Old Left appears less radical today than formerly. It would have supplanted the managerial elite of capitalism with one of its own under state control. Allegedly, this change would have oriented productive and distributive systems to the public good. This essay cannot deal with the various difficulties older socialist and communist countries have faced in this enterprise—the inefficiencies of central planning, the new inequities that have arisen in the distributive system, the military orientation of national budgets, the bleak effects of state management on the quality and attractiveness of goods, and the informal reappearance of somewhat inefficient cost–price markets in countries that tried to do without them. More germane to our discussions: the Old Left has proved far from radical on the subject of large-scale organizations. It left them intact. If anything, the left has exacerbated their defects. It has generally encouraged single-purpose institutions to remain fixed upon a primary task to the point of rigidity. It has elevated hierarchical principles of organization to a quasireligious status. It has flattened out performance and banished the personal from the public arena. No wonder that the historian of political thought, Sheldon Wolin, in seeking to describe the overarching theme in modern political thought, turned to the phrase, the "Age of Organization."[15] Although East and West differ in substance, they share the form of the large-scale institution.

From this perspective, the New Left made a truly radical move. For a feverish interlude, it wanted to dismantle the bureaucracies. Huge, single-purpose institutions had contributed to a destructive and doomed system. They could not innovate or invent; they could not respond flexibly.[16] Their vaunted productivity had blighted the environment. Their hierarchies oppress; their officialdom entraps; their horizons press down and narrow. Their standardized products, from large automobiles to fast and fatty foods, endanger health and survival. These and other charges have spilled out of books and TV talk shows that huge corporations fund and control. The fever of the 60s has abated, but an appealing echo of it reverberates in those who would lower expectations and scale down institutions to a more modest human compass.

Small may be beautiful, as humanists from Jefferson to Schumacher have reminded us; but not always, and not in all things. The United States may, in fact, be headed toward an era of duplex social organizations, including both large, geometrical, relatively single-purpose institutions and small, informal, somewhat more spontaneous communities that both supplement, counterpoint, and criticize the first type of organizations, but also experiment with new forms proleptic for the future. (Herbert Read, in the *Grass Roots of Art,* notes that Egyptian civilization, at an important stage of its development, produced two kinds of art—one the formal, geometrical art of the pyramids, and the other, a more naturalistic, lyrical, spontaneous tradition of craft arts.) The dialectical relationship between these two types of social organization already shows up in human services. In education, health care, and the care of the elderly and retarded, one needs not only the larger institutions that can fund and organize the talents of professionals, but also smaller-scale groups—the churches and other voluntary communities—that can mobilize amateurs to serve human need. A similar mixture of large and small institutions has begun to reappear in commerce today.

The Regulators and Reformers

The coexistence in our society of smaller organizations and corporate giants hardly solves the problem of regulating or reforming the latter. American society has more often tried to regulate corporations than to reform them. Society uses as its chief regulatory systems, the marketplace and the law. Although the marketplace appears to be the arena of freedom and initiative in contrast to the world of government restraints, the marketplace also regulates. As Lester Thurow has pointed out in the *Zero Sum Society,* the marketplace relies on a very elaborate set of rules and agreed-upon signals and government sanctions in the course of its operation. "Without government regulations, there are no property rights and without property rights there is no free market."[17] One chooses not between two systems, unregulated and regulated, but between right and wrong regulations. The marketplace as a major and largely positive force in directing corporations to social responsiveness and responsibility cannot do the job alone. The give and take of the buyers and sellers in the marketplace arena does not, in and of itself, effectively curb corporate social injury.

> . . . those who have faith that profit orientation is an adequate guarantee of corporations realizing socially desirable consumer goals are implicitly assuming: (1) that the persons who are going to withdraw patronage know *the fact* that they are being "injured" (where injury refers to a whole range of possible grievances, from getting a worse deal than might be gotten elsewhere, to purchasing a product that is defective or below warranted standards, to getting something that produces actual physical injury); (2) that they know

> *where* to apply pressure of some sort; (3) that their pressure will be *translated* into warranted changes in the institution's behavior. None of these assumptions is particularly well-founded.[18]

The law, through a set of discrete prohibitions, disincentives, and constraints, provides a second resource for curbing socially irresponsible corporate behavior. No agency other than the federal government and no instrument other than federal law can match the power of corporations. In some cases, these corporations exceed in economic power all but the largest of nations. Not until recently have the courts begun to require of the corporations a responsibility commensurate with their privilege as legal "persons." But Stone and other critics have doubted the ability of the law to produce corporate social responsibility. The escape hatch of limited liability, the frustrating diffusion of managerial responsibility, the reactive rather than the preventive nature of the law, and myriad difficulties in making and enforcing laws make some critics look elsewhere for help. Quite apart from its difficulties in detail, the strategy of relying on external regulation alone falters at a crucial point. It generates a latent antinomianism—a barely repressed spirit of lawlessness—in the regulated. Regulations usually need to enforce minimal standards. But the regulated turn these standards upside down and use them as guidelines to mark out for themselves the maximum of what they can get away with. Thus, critics like Peter Drucker and Christopher Stone have looked more to the reform of corporate structure than to discrete, substantive regulations as the route to corporate social responsibility.

Reformers fall into two patterns. One group would decentralize the corporation in order to distribute responsibility more widely. Peter Drucker, for example, identifies four basic designs, above and beyond the traditional, hierarchical pattern: the task-oriented team (which the health care team in a hospital or a research secretariat in a large organization illustrate); the decentralized but federated company (of which GM serves as the prototype); a simulated decentralization (Monsanto and IBM) in which different functions within a whole organization disperse into relatively autonomous companies; and a systems design in which a variety of institutions, public and private, collaborate in producing a result (NASA).[19] Quite apart from the special merits and defects of each design, none widely distributes the responsibilities of top management. Chevrolet, a decentralized unit in GM, still, as an enormous organization, puts power in the palms of a handful. (The task-oriented team comes closest to distributing responsibility, but it usually merely advises top management.) Whatever the design, top management still dominates. As Drucker himself admits, only four or five people share in the uppermost levels of power.

More boldly, some experiments would include workers in decision-making. This move toward industrial democracy would patently serve to emphasize the well-being of the workforce as a central goal of management. Skeptics worry, however, whether the design would provide the flexibility and efficiency that survival in a semicompetitive economy requires. Others interested in corporate social responsibility question whether analysts assume naively that large groups of essentially self-interested employees will respond more sensitively to issues of social conscience than do the traditional handful of self-interested top managers. To build social conscience into the enterprise requires other kinds of shifts in structure and ethos.

A second group of reformers has turned away from efforts to decentralize authority and sought instead to increase responsibility at the uppermost level—the board of directors. To this end, Christopher Stone would provide the board of directors of a corporation with its own staff capability to keep the board from being a rubber stamp for top management. More controversially, Stone would assign a sliding percentage of public directors to represent the public interest on the boards of our largest corporations. In companies with special problems—pollution, product safety, and so on—additional such special directors would strengthen its conscience and competence to make the relevant decisions. Stone's reforms spread beyond the board. He also recommends various changes in the managerial structure to increase the flow of information upward and downward in the corporation. Too many managers get off the hook by pleading ignorance. Increased knowledge will enforce increased responsibility.

The rationale behind these reforms suggests that the corporation needs to develop a more interior sense of responsibility than the external threats of the law or pressures of the market will develop. We can frame no better image of interiority than the structure of knowledge, conscience, and responsibility in the human person. Thus Stone and others[20] take advantage of the legal status of the corporation as a person and try to build the analogous ingredients of personal knowledge and conscience into its life. Even more specifically, Stone develops a Freudian model. He discerns a structural analogy between the functions of the ego, superego, and id in personal life and the place of top management, the board, and the profit motive in the corporation. The source of rational decision-making in personal life—the ego—corresponds to the managerial staff. The unconscious drives, libidinous and otherwise—the id—correspond to the profit and other motives that make managers tick. Finally, and most crucially, the voice of conscience in personal life that internalizes the dictates and demands of parents in particular and the society at large—the superego—finds its corporate analog in the board of directors. Hence the importance of board reforms in Stone's proposals.

Formally, Stone's work suffers from inconsistency. He works hard to show why the "law can't do it," why external threat and sanction do not work. And yet, he himself urges that the law mandate far-reaching changes in the corporate structures. The law can't do it; the law must do it! He thinks, of course, not that "the law can't do it," but that discrete, substantive laws cannot do it. He proposes instead even more invasive laws that would reform the corporate structure.

The compulsory legal ingredient in his proposals leads to a second problem. Public directors, staff trouble-shooters, and ombudsmen imposed on the corporation from without will hardly interiorize responsibility. They will form an isolated and stigmatized in-house KGB unless other deeper transformations of the corporate ethos proceed simultaneously. Finally, Stone quite rightly insists that the corporation needs a better "information net." Bad news, for several reasons, tends not to flow upward. Underlings do not want to bear it and authorities do not want to hear it. Neither wants his or her innocence compromised. But the flow of knowledge—not just technical information, but moral knowledge and judgment—will not flow within the corporation at large unless the corporation rethinks the manager's role as leader and teacher.

Proposals for Transformation: Managers, Consultants, and Middle Managers

The very word, "management," emphasizes a custodial function—the manager handles a volume of business by invoking rational, impersonal routines. But no corporation can afford to reduce itself wholly to routine. The business corporation has to deal with the market—which forces its managers daily and relentlessly to deal with uncertainty—the uncertainty of customers, raw materials, transportation, and energy costs, changes in taste and habit. Thus some analysts would reserve the term "bureaucracy" for nonprofit service organizations or public institutions, arguing that any organization, large-scale or otherwise, working in the market, must maintain its flexibility, its adaptability, if it would survive. Indeed, the effort of a bureaucracy to eliminate uncertainty through routine produces its own uncertainties. The effort to eliminate risk itself risks creating rigid, inflexible, interminable procedures. Business management should not eliminate risk, but figure out the right risks to take.

Thus management seeks to interpret itself less in custodial terms than in the language of leadership. The term "leadership" invokes etymologically the notion of a journey. Going where? Into the unknown. Into the not-yet-fully-revealed. Into the X of the future. Thus seminars on leadership invariably link leadership with the decade ahead: "Leadership and the 80s." Leadership connects the present to the future, a future relatively opaque, partly self-determined and partly determined by forces and factors beyond one's control.

Leadership consists of what the German philosopher, Martin Heidegger called *"Vorlaufen"* (literally, running ahead of oneself; less literally, anticipation). Anticipation in our specific context includes several ingredients: first, the selection of goals—resolving upon a destination; second, determining the route to these goals; and third, as the Old English root for the word "leadership" suggests, causing others to go by showing the way. In business, top management assumes these tasks are its chief responsibilities.

Leadership requires facing the uncertainty of the future, the ultimate uncertainty of which, Heidegger felt, no rational structure can wholly eliminate or ignore. But, in the everyday world of practice, leaders have sought to reduce it to a minimum. Since time immemorial, those charged with great responsibilities have sought to reduce the element of risk by rendering the future as transparent as possible. Thus the Greek general, charged with the burden of decision-making, had recourse to the oracle who read the entrails of birds to discern the future, to convert its X into something more knowable and psychologically manageable. The revolutionary leaders of the 19th century had similar recourse to prophetic judgment to escape mystery. Karl Marx prevailed over other socialist traditions primarily because he placed revolutionary action within the context of a philosophy of history. He reduced the X of fate into the iron rule of historical necessity. Thus those very events—war, depression, famine, and market collapse—that previously convinced the proletariat of its powerlessness, Marx called signs of the imminent collapse of capitalist society and the triumph of the proletariat.

The outside consultant or advisor to top management provides a smaller service to 20th century corporate society. No less prey to anxiety before fate than the Greek general or the European revolutionary, the manager today appeals to the outside counselor/oracle. Consulting has become, like psychiatry, a modern growth industry. It offers moonlighting to academicians and handsome careers to other professionals. Presumably, the modern advisor/consultant offers something better than the eviscerated entrails of birds: it offers *harder* data. In practice, though, the modern expert has devised ways to hedge the bets. In earlier, more epistemologically optimistic days, the expert talked about predictions; later, and more modestly, the same expert offered projections, and still more recently, scenarios, to help protect himself against the vagaries of recommended outcomes.

The consultant purports to sell knowledge, largely technical knowledge which helps the client know better how to get from here to there, but also partly substantive, critical, or moral knowledge that raises the question of whether the there is worth getting to. Especially the latter kind of knowledge functions in a peculiar way with respect to uncertainty. It does not wholly eliminate uncertainty and risk; at its best, it helps determine what risks to take, what uncertainties to bear.

The consultant offers, further, indispensable psychological services that supplement his or her function as critic. The very nature of managerial work produces myopia. Managers draw up their daily round of responsibilities close to their eyes. Managers lose psychic distance and watch only where they plant their feet and not where the road leads. Managers also lose critical perspective on their colleagues and the advice they offer. Thus the outside consultant provides them with a little space in which to maneuver, a little discretionary power lost in routine, a chance to see afresh a world grown overly familiar. The consultant helps the manager keep professional distance. The quality of detachment distinguishes the professional from the amateur. Physicians do not take care of their own families because they would yield to emotion and could not offer the most effective help to their own people. Similarly, the corporation draws the manager so completely into its inner life and rhythms that he or she risks losing the professional eye of a stranger. Ideally, professional detachment allows for the possibility of moral transformation; it offers the space and distance required to move in a new direction.[21]

Finally, the consultant provides top managers with some of the benefits of collegiality that a hierarchical organization often denies them. Isolation on top of the heap forces managers to look outside for uninhibited discussions of issues difficult to raise within. In a more sinister role, the outside consultant protects community within the organization, by serving as convenient scapegoat. The manager can blame the consultant for the more unpopular innovations. The consultant draws lightning away from those who have to continue to work within the company's confines.

Within the corporation, nevertheless, a new breed of middle managers has increased rapidly to provide some of the services that the outside consultant offers. These knowledge experts work on discrete tasks and offer advice to top management. They are called "middle managers" in a sense, a misnomer. They do not take responsibility for the work of others who report to them. They work in staff rather than in line positions. These advisory "middle managers" are next in the hierarchy, but they work somewhat more professionally and collegially. They derive authority less from position than from knowledge; they rely less on command and obedience than upon research, reflection, and persuasion.

Like the 18th century *philosophes,* these middle managers exercise power by whispering in the ear of the powerful. Inevitably their power suffers two limits. First, it depends upon enlightenment (or a readiness for enlightenment) at the top. The king must wish to learn. Seldom can he be outvoted. Second, the king may restrict the advice he wants to the purely technical—in which case, the knowledge expert or professional misses whatever opportunities to transform that access to power offers. These managers have already been trained and housebroken in such a way as to reduce professional responsibility to technical competence alone. In effect, they say to the boss: "tell us what your goals are and we will tell you whether their attainment is feasible—as lawyers, as engineers, as accountants, or as management experts."

A different opportunity, of course, presents itself to the middle manager/adviser if the very definition of the managerial task includes critical judgment about goals and not just tactical cunning in attaining them, if it carries with it some obligation for moral substance, and not just instrumental intelligence. Those *ifs* imply a somewhat more transformational rather than transactional understanding of business leadership.

Transformational Leadership

A contractualist ethic of the marketplace tends to produce a transactional understanding of leadership. The leader in a transactional context (whether functioning as a professional, a manager, or a politician) takes at face value the wishes and interest of people and offers services that gratify explicit aspirations for power, money, relief from distress. Thus patients and clients purchase from professionals help in overcoming troubles that hinder clients in the pursuit of their own goals. Political leaders in the social contract state acquire power by gratifying the wants and interests of their citizens. Managers contract to owners to maximize profits by satisfying in turn the wants of customers. All of these transactions amplify one's powers, but leave unaltered one's character and commitments. Wishes, needs, interests escape untouched, untransformed; managers basically gratify rather than change them.

A more covenantal understanding of institutional and professional responsibility entails a more transformational understanding of leadership. James MacGregor Burns particularly has seen the need for this sort of leadership in the political order. Such leadership entails helping people reorder their wants and desires, not just gratifying them. It means lifting them out of their private, everyday lives into a more spacious community of public purpose. Burns writes in his book on leadership: "That people can be lifted *into* their better selves is the secret of transforming leadership and the moral and political theme of this work."[22] A society of expanding frontiers and gross national product, with an abundant supply of energy and places to dump waste, does not notice its need to transform. Such a society can waste—indulging wants and desires because growth will gratify the surplus of unmet wants in due course. But a minimum growth democracy like ours today must choose one of two courses. Either it develops into a veto-profuse society in which each power group protects its possessions through its very considerable defensive capability. In this case, no serious, coherent program to solve any major problem can prevail; one interest group or another nibbles each part of a program to death. The society is paralyzed. Alternatively, the society finds some way of reordering its priorities and habits in order to address, in some imperfect but meliorative way, its fundamental problems. The latter is the task of transformative political leadership.

Perhaps in reaction to earlier patterns of paternalism, transactional leadership has until recently prevailed in the professions. In the practice of medicine, transactionalists have assumed that the improvement of health depends on the allocation of an ever-increasing portion of the GNP to medical research and care. But clearly, improvements in health today depend more on changes in personal habits and the quality of environment than they do on expanding and expensive piecework, marketplace medicine. Preventive medicine today transforms habits and structures. The point carries over into other professions. The emphasis in the current legal *Code of Professional Responsibility* falls on the lawyer as advocate who contends with crisis in the courtroom. But counseling supplies a functional equivalent to preventive medicine in the practice of law. The proposed new *Model Rules of Conduct* recognize the importance of this preventive function by placing the lawyer's duties as counselor first on their list of responsibilities. Further, the proposed code expands the lawyer's advisory duties to include not only technical information on the law, but also discussion of the moral features of the case.[23]

The act of anticipation is the functional equivalent of preventive medicine and legal counseling for the business leader if he or she does more than manage crises. Business schools have, of course, trained transactional leaders, adept managers of quandaries and crises, rather than leaders who transform the ethos and structures of their companies. The prevailing case method of study in business schools tends to be reactive rather than proactive, improvisatory rather than anticipatory. It directs leadership insufficiently to transformation.

On the whole, transformational leadership has not won a good reputation in the recent life of this country. We have seen good cause to worry about overbearing efforts at transformations that have ended up deforming. We recall a Puritan past that declined into moral officiousness and a cramping paternalism. Many of us have suffered therapists who have ignorantly preached their politics when we would have preferred to pay them for their professional services and let them be on their way. Efforts to transform, moreover, founder on pluralism. A society with diverse constituencies can ill afford too many contrary efforts to transform. So argues the libertarian who would restrict instrumental goals. Why not assume that community among us serves merely useful rather than moral ends, that it gratifies our wants rather than fosters in us the aspiration to excellence? Why not rely on mechanisms rather than insight to cope with conflicting interests?

THE PROFESSIONAL/MANAGER AS TEACHER Transformational leadership suffers the aforementioned difficulties unless one includes teaching as its essential ingredient in a pluralist society. Such teaching attempts to lay out and clarify an agenda for action. Leadership that teaches does not attempt to transform people by bending them against their will, or dazzling them out of their faculties, or managing them behind their backs. It does not simply issue a blind call for a particular decision or strive to indoctrinate without illuminating. Rather, it opens up a horizon in which clients and colleagues see a given world of practice in a new light and therefore unlocks a freedom to perform in new ways.

Unfortunately, professionals have vastly underestimated the therapeutic value of teaching in professional life. The quarrel on this issue in medicine is very old. It goes back to the "rough empirics" of classical Greece who used to ridicule their more scientifically oriented colleagues for attempting to teach their patients.[24] Physicians, however, must teach their patients if they would enlist them more actively in their own health maintenance or rehabilitation. P. Lain Entralgo, the Spanish historian of medicine, argued that words are to drugs what a preamble is to a constitution. They help provide the clarifying context for the therapeutic regimen. Other professions should undertake a similar teaching obligation. Lawyers offer technical services in drawing up contracts and appearing before the Bar, but they must also teach their clients as counselors if they would keep them out of the courts. Accountants organize data according to the technical canons of the profession, but they also provide an interpretive framework for decision-making. Politicians, public administrators, managers, and middle managers offer a kind of professional expertise, but preeminently, as transformative leaders, they must teach. The constitutional powers of the presidency founder if the President of the United States fails in that office of which the Constitution never speaks: as teacher to the nation. Similarly, managers must teach if they would "show the way" and lead by persuasion.

(1) The top manager heretofore has tended to control his or her world by command within and by manipulation and pressuring without. But managers have greatly exaggerated the necessity of inscrutability within and secretiveness without; they overlook the cost of these two tactics. When one governs within exclusively by command rather than persuasion, one controls behavior rather than leads people and one fails to prepare another generation for leadership. Further, coping with the outside world exclusively by manipulation (advertising) and bullying (lobbying) generates distrust. False claims and threats produce skepticism and resentment. Neither the customer nor the government listens to the merits of the case. Top management should accept a teaching responsibility both without and within.

It would be foolish, of course to exaggerate in these matters. No large-scale institution, public or private, can wholly substitute persuasion for command. The presidency may offer a bully blackboard to the nation, but no President can afford to restrict official decisions to the boundaries of what can be explained or defended by persuasion. The board meeting of a corporation or a committee meeting of its executives hardly resembles a seminar. The board must close and make decisions long before everyone has understood or yielded to persuasion. But the "information net" of the corporation will not be mended in other than superficial ways unless substantive issues, as well as technical intelligence, can surface. Top management can signal its interest in matters substantive.

Upper level managers accept implicitly a further teaching responsibility of a less public nature: they must act as mentors in bringing along their junior colleagues. In some respects, Americans may not address this obligation as wisely as business leaders in other countries. In America, the boss or sponsor, more often than not, serves as mentor to subordinates. The complication of roles skews, and sometimes corrupts, the relationship. The Japanese have spared themselves this difficulty by designating as mentors senior persons who do not exercise direct line authority over their junior colleagues.[25] The Japanese thus avoid the dangers of parentalism with avuncularism. The French system differs from both the American and the Japanese systems. It encourages mentoring among equals. Workers do not compete against one another directly for promotion through performance. Promotion comes through external examinations.[26] The French system, whatever its other defects, reduces face-to-face competitiveness and encourages more solidarity and mutual support on the office floor among colleagues at the same level. However accomplished, mentoring contributes to building responsibility throughout the corporate culture.

(2) Junior colleagues, especially in staff positions, accept implicitly a corresponding obligation to teach their superiors on policy issues in their role as advisors. Such teaching, of course, in a purely contractualist setting means transmitting information that the boss can use instrumentally to reach pre-established goals. A great deal of teaching in professional life does just that, and should do that. But in a more inclusive covenantal setting, the advisor needs some space to raise substantive moral questions about goals. Otherwise, the very meaning and social justification for the professional, the knowledge expert, diminishes. Presumably, the professional generates power through knowledge placed at the service of human need. It would be odd so to restrict the scope of this service in the corporate setting as to provide the middle manager with less moral power than Balaam's ass, who posed an awkward question or two about the direction in which his master rode, or with less freedom to contest, argue, and explain than the maid-servants in Moliere's comedies.

Interior transformations of corporate life will not eliminate the need for radical departures or for external regulations and reforms. Radical experiments must supplement the institutions currently in place to develop those forms that may be proleptic for generations to come. Regulations can shift incentives; and reforms can orient huge institutions toward their proper ends. But regulations and reforms without more interior changes will founder. The opportunist in us all dodges nimbly and will find a hundred ways around regulations and through reforms if the basic perception of the professional or manager's vocation does not change.

Notes and References

[1]A 1963 issue of *Daedalus*, devoted to the professions, euphorically projected a limitless demand for professionals that could not possibly be met in this century. Within twelve years, all professions with the exception of medicine had experienced some oversupply.

[2]Geoffrey C. Hazard, Jr., *Ethics in the Practice of the Law* (New Haven: Yale University Press, 1978), p. 7.

[3]Abraham Flexner, "Is Social Work a Profession?," *Proceedings of the National Conference of Charities and Corrections* (Chicago, 1915).

[4]W. Metzger, "What is a Professional?," *Seminar Reports,* Program of General and Continuing Education in the Humanities, Vol. 3, No. 1 (New York: Columbia University, 1975).

[5]Peter F. Drucker, *Management: Tasks, Responsibilities, Practices* (New York: Harper and Row, 1974), pp. 551–559.

[6]Milton Friedman, "The Social Responsibility of Business Is to Increase Profits," *The New York Times Magazine* (Sept. 13, 1970), pp. 32–33.

[7]Fred Luthans and Richard M. Hodgetts, *Social Issues in Business* (New York: Macmillan, 1976), p. 40.

[8]Keith Davis outlines the tie-breaker argument in "The Case for and Against Business Assumption of Social Responsibilities," Luthans and Hodgetts, *ibid.*, p. 96.

[9]John G. Simon, Charles W. Powers, Jon P. Gunneman, *The Ethical Investor* (New Haven: Yale University Press, 1972), pp. 15–21.

[10]Robert Hay and Ed Gray use developmental language about the fundamental commitments of business; they refer to "historic phases." But then they admit that "each new phase did not replace the earlier phase but rather superimposed on it." See their "Social Responsibility of Business Managers" in Luthans and Hodgetts, *Social Issues,* p. 110.

[11]Max Weber, *From Max Weber: Essays in Sociology,* H. H. Gerth and C. Wright Mills, eds. (New York: Oxford University Press, 1958), pp. 214–216, 295–296.

[12]Sheldon Wolin, *Politics and Vision* (Boston: Little, Brown, 1960), pp. 379–380.

[13]American Bar Association, *Code of Professional Responsibility and Code of Judicial Conduct* (1978).

[14]Lon L. Fuller, *The Morality of Law* (Clinton, MA: The Colonial Press, 1964), pp. 3–32.

[15]Sheldon Wolin, "The Age of Organization and the Sublimation of Politics," *ibid.,* Chapter 10.

[16]The Berkeley student leader complains, "As bureaucrat, an administrator believes that nothing new happens. He occupies an a-historical viewpoint." See Mario Savio, "An End to History," in *The New Student Left,* Mitchell Cohen and Dennis Hale, eds. (Boston: Beacon Press, 1967), p. 248.

[17]Lester Thurow, *The Zero Sum Society* (New York: Basic Books, 1980), p. 129.

[18]Christopher Stone, *Where the Law Ends: The Social Control of Corporate Behavior* (New York: Harper and Row, 1975), p. 89.

[19]Peter F. Drucker, *Management.*

[20]Kenneth Goodpaster, formerly of the University of Notre Dame and now at Harvard University, joins Christopher Stone in the attempt to model the corporation after the structure of moral agency in a person. In this respect, they differ from the philosopher, John Ladd, who operates with a mechanistic model: the corporation as a rational machine. See Ladd's essay, "Morality and the Ideal of Rationality in Formal Organizations," 54 *Monist,* no. 4 (October 1970): 488–516.

[21]For more on the place of anticipation and the need for space in corporate decision-making, see Charles W. Powers and David Vogel, "What is Business Ethics?" in *Ethics in the Education of Business Managers* (Hastings-on-Hudson, NY: The Hastings Center, 1980), Chapter I.

[22]James MacGregor Burns, *Leadership* (New York: Harper and Row, 1978), p. 462.

[23]Rule 2.2 of the proposed *Model Rules of Professional Conduct* breads: ''Narrowly technical legal terms may be of little value to a client . . . It is proper for a lawyer to refer to relevant moral and ethical considerations in giving advice. Although a lawyer is not a moral advisor as such, moral and ethical considerations impinge upon most legal questions and may decisively influence how the law will be applied.'' Discussion draft prepared by the Commission on Evaluation of Professional Standards of the American Bar Association, January 30, 1980.

[24]For an expansion of this insight in the classical Greek context, see P. Lain Entralgo, *Doctor and Patient* (New York: McGraw-Hill, 1969), pp. 29–41.

[25]For the Japanese system of mentoring, see Peter F. Drucker, *Management,* pp. 254–257.

[26]For the French system of mentoring, see Michel Crozier, *The Bureaucratic Phenomenon* (Chicago: The University of Chicago Press, 1964).

Accountability and the Bureaucratization of Corporations

Thomas Donaldson

"Man is born free, but everywhere he is in organizations." Whoever thus altered Rousseau's original dictum must have been aware of the facts of life in large, modern corporations. Large corporations are succumbing to the pressures of bureaucratization, and experiencing the problems typical in bureaucracies. Their transition carries direct implications for the issue of accountability, because the extent to which people are submerged in and controlled by bureaucracies is the extent to which ordinary individual accountability is diminished. The aim of this paper is to examine accountability in the large modern corporation, and to show how the increasing bureaucratization of corporate life threatens the very possibility of meaningful moral accountability.

As organizations increase in size, they become more bureaucratic; and modern corporations are clearly becoming larger. The same share of manufacturing assets controlled by the largest 150 corporations in 1980 was controlled by the largest 1000 corporations in 1946. More important, changes in basic structure are occurring; corporations look less like the traditional Weberian ideal, with clearly defined authority and accountability structures, and more like complicated aggregates of impersonal rules, centralized decision-making networks, and isolated authority strata. The demands of technology have forced the development of a corporatre technostructure and have blurred, in turn, traditional demarcations of authority and accountability. We shall see that the resulting problems of accountability are logical as well as empirical. There are logical difficulties in assigning accountability to corporate structures that divorce responsibility to account for moral error from the capacity to control events.

Even determining the locus of accountability is more difficult in a bureaucratic corporation. Contrast the ease with which accountability may be understood in a small, simple organization, with the difficulty of understanding it in an enormous one. How, for example, do we sort through the tangled skein of accountability problems that culminated in the Ford Pinto disaster?

For present purposes, let us understand "accountability" to mean the responsibility a person or group has to others for providing an account of past, present, or future behavior, where (1) the account assumes standards of evaluation shared by those receiving the account, and (2) the assignment of praise or blame is appropriate depending on the degree of congruence between standards and behavior. This is a rough definition, but it handles standard cases of, for example, the sea captain, accountable for the conduct of the ship, and the lawyer, accountable for the client's defense. Three specific tendencies constitute the overall movement toward bureaucratization: (1) the increase of impersonal rules; (2) the move toward centralized decision-making; and (3) the isolation of strata in the corporate hierarchy.[1] We shall examine each separately.

The subjugation of the individual by the organization is an old fact. While examining the modern corporation, it is well to remember that only two centuries ago, conformity in organizations was obtained through very direct means, often with a great deal of open coercion. Nothing less than complete devotion was demanded of members of the Jesuit order, the Prussian Grenadiers, or the employees of the Frugger House. In such organizations, leaving was equated with treason.[2] Yet modern methods of generating conformity have an equally effective, though more subtle impact. Of special importance is the ongoing deterioration of systems of direct supervision in favor of elaborate systems of impersonal rules and regulations. Although people are relieved from the watchful eyes of their superiors when their work becomes rule-bound, special problems follow for morale and accountability.

Organizational theorists agree that increasing bureaucratization of the corporation results in an increase in impersonal rules. Max Weber maintains that the evolution towards large-scale organizations is unrelenting, and he identifies an increase in rules as a necessary feature of that evolution. In order to achieve increasing efficiency, expertness, rationality, and predictability, he asserts that organizations must develop effective control structures, specify spheres of competence, and, in turn, increase the number and the impact of rules.[3] In an efficient organization, individual people must be replaceable without provoking crisis, and this means that decision-making must depend on rules, not people. Other forces also prompt systems of impersonal rules. The subordination of one individual to another creates a predictable tension (especially in societies where libertarian ideals are strong) and these tensions can be alleviated by imposing impersonal bureaucratic standards. But a vicious circle develops. Impersonal rules perpetuate the very tensions that generated them: impersonal rules reinforce low motivation, and this, in turn, creates a need for close supervision.[4]

The immediate consequence of the emergence of impersonal rules is that accountability becomes submerged in rules. We noticed that a special advantage of rule structures is that they are more permanent than people; they relieve the organization from a dependence on particular individuals. Yet from the standpoint of accountability this advantage becomes a disadvantage in that rule-bound individuals refuse accountability for their own actions. "I only follow the rules," is the typical, threadbare, bureaucratic response. If the antagonisms between worker and manager are severe, the curious phenomenon of ritualism can develop. We are all familiar with the stubborn employee who makes a point of following his regulations to the letter, even when doing so involves ignoring realities and frustrating the very goals of the organization. Members of a labor union may, in a similar manner, defy management by "working to rule"; that is, only working up to the level explicitly stated in the union contract, *even when* exceeding that level is easier.

If accountability is submerged in rules, then it follows that ultimate accountability attaches to those who make the rules. But though this implication is logical, it neither simplifies nor resolves certain problems. Rules outlive their makers, and it is often impossible to hold a single person accountable for a bad rule (or for the exceptions an otherwise good rule should allow). Furthermore, in a bureaucracy individual people seldom make rules alone: committees or informal groups typically generate rules, with the result that these groups become the logical loci for accountability. This poses obvious problems, however, since the committee or group is at a distance from the clerk or employee who follows its directives, and since such groups can only *account* for their activities when they are in session.

In addition to impersonal rules, corporate bureaucracies generate centralized authority. As Paul Kurtz remarks, "The logic of the organization is essentially conservative. Thus there is a standardization and consistency of behavior. Increasingly there is a tendency for individual responsibility to give way to corporate responsibility, and the individual denies he is responsible for what the corporation does."[5] The elimination of discretionary personal power in lower corporate ranks pushes that power up the ranks. Power tends to flow from the center of the bureaucracy to its edges, and when the bureaucracy is large, the lines of accountability become overextended. John Lachs characterizes the resulting problem as one of "psychic distance": When the Japanese General Yamashita was tried (and eventually executed) for war crimes following World War II, he protested that the atrocities his soldiers committed in the Pacific Islands were so distant from the center of his organization that they occurred despite his own good intentions. Lachs observes that the centralization inherent in large bureaucracies demands that accountability be assigned to the center; but this assignment is weakened by the fact that the psychic distance between center and periphery is often so great that effective *control* vanishes.[6]

Closely connected to the problems of centralization and impersonal rules is the problem of the isolation of different strata in the corporate hierarchy. When authority is converted into impersonal rules and when ultimate power is transferred to the center of the corporation, the result is a separation of the strata in the authority hierarchy. Impersonal rules obviate the need for face-to-face authority relations, and this, in turn, means a separation of subordinate and superordinate strata. One always obeys the rules, but it no longer is necessary to submit to the whims of individual people. One is separated from the next higher rung on the authority ladder. Here, without the need to yield to higher authority, the importance of peer pressure increases. The peer pressures to which such people submit, however, are typically not, either by accident or design, ones that assume accountability for the actions of the organization as a whole. Peer pressures and impersonal rules eliminate day-to-day decision-making in a corporate bureaucracy, but both discourage genuine accountability.

Because of the isolation of the various strata, it often happens that the center of the organization, i.e., its decision-making nucleus, is isolated from the peripheral strata at which the organization has its direct contact with the public. Here the problem of isolation of strata overlaps with that of centralization. Executives at the center are unable to respond effectively to, and be accountable for, actions at the periphery. In his classical analysis of General Motors in 1946, Peter Drucker identified one of General Motors' greatest problems as the isolation of its top executives from the sentiments of the general public—an isolation that resulted in serious bureaucratic blunders.[7]

We have seen that because a bureaucratic organization discourages initiative at the periphery, decisions must be made, and accountability must reside, where the power is located, at the center. There is one exception to this rule. Modern corporations require one kind of employee whose actions stubbornly resist being reduced to impersonal rules—the professional. The professional is an expert whose services are required because of increasing technological demands or increasing complexity in the surrounding social and legal climate. Yet as an expert the professional is presumed to know best about his or her area of work, and this is unique in the bureaucratic world. Professional activities cannot be fully directed from the main office through impersonal rules, because only the professional is qualified to direct that work.

Two important consequences follow. First, the professional, unlike other bureaucratic employees, is faced with two sets of standards used to evaluate his or her behavior; and the two are not always mutually compatible. Professionals are faced on the one hand with the standards of the organization that dictate success in terms of organizational goals; and on the other hand, they are faced with the standards of their professions. The professional feels this dilemma most acutely when thrown into the role of organizational administrator. The professor who becomes dean of the college suffers when forced to make important decisions that supposedly *both* enhance the future of the university and satisfy professional academic standards. The closer the professional comes to the power center of the organization, the sharper the conflict.

Second, in a bureaucratic world that tends to reduce all authority relationships to impersonal rules, the relative status and power of the professional increases. "The position of experts," Crozier remarks, "is much stronger in an organization where everything is controlled and regulated,"[8] so long, of course, as the professional's own task cannot be reduced to rules. It follows that as corporations become more bureacuratic, the relative power of the professional increases in relation to that of the nonprofessional. The bureaucracy transforms specialized expertise into political power.

In addition to the traditional categories of professionals, modern corporate life creates new ones. The data systems analyst, the marketing specialist, the labor negotiator, the management theorist, and the public relations expert are necessary ingredients in the modern corporate success formula. These new professionals also possess most of the traditional characteristics associated with professionals: they rely on a theoretical store of knowledge, are graduated from research-oriented institutions, apply their knowledge to practical problems, and subject their work to review and criticism from colleagues.

Many of these new "technocratic" professions, however, lack a key characteristic associated with traditional professions. With the professions of medicine, law, or teaching, we associate some spirit of altruism or service; but the new technocratic professions often lack this characteristic, and thus create problems of accountability. We associate the goal of healing with the physician, of knowledge with the professor (no matter how mercenary doctors or professors may in fact be), yet there are no corresponding goals for the marketing specialist, the public relations manager, or the advertising expert. The professional standards of modern technocracies do not include standards of moral accountability in part because the standards do not even refer to an altruistic element in the overall goals of their respective professions. The old professions have frequently failed to embody adequate moral standards in order to make their practitioners *professionally* accountable for moral behavior; but the new professions are doomed because they do not even attempt it.

The fact that the power and status of professionals rises in proportion to the degree of bureaucratization of the organization and the degree of technological efficiency required, means that the need for moral accountability *through* one's profession increases. Henry Ford used to make virtually every decision about the design and production of the Model "T." Today the demands of technology reserve authority for the technocrat. At Ford Motor Company important decisions are reserved for the production specialist and the design engineer, among others. But the added authority for the new professional requires added accountability: those who design and produce Ford products today are obliged to avoid catastrophies such as the Pinto's exploding gas tank. That is why in the modern corporation the need for professional ethics is so very pressing.

When Peter Drucker returned to General Motors nearly twenty years after his original study, he complained that GM had failed to solve the basic problems of balancing its own needs with "concern for its environment and compassion for its community."[9] Drucker concludes that the blame for this failure lies at the door of the new professional. "General Motors' success is clearly the success of the technocrat," he remarks, "but so is General Motors' failure."[10]

So far we have identified four fundamental accountability problems stemming from the increasing bureaucratization of the corporation. Again, these are related to: (1) the increase of impersonal rules; (2) the increase of centralized decision-making; (3) the increase of isolation between hierarchical strata; and (4) the concomitant increase in the relative power of professionals and a failure, especially of new professions, to embody standards of moral accountability. It is tempting to turn to traditional answers, namely, to the power of government, or of union advocacy. But the government finds it difficult to control the professional for the same reason the corporation finds it difficult: the acknowledged expert in the professional's area is in fact another professional. And the unions are more concerned with the welfare of their members than they are with consumers or society at large. A final consideration is still more damning. How can the accountability problems stemming from the bureaucratization of the corporation be solved by turning to the government or the unions—which are *themselves* large bureaucracies?

Society demands that corporations be morally accountable for their actions. That is to say, society demands that corporations be responsible to consumers and the general public for providing an account of their behavior in accordance with normative standards of, e.g., product safety, use of natural resources, hiring policies, environmental impact, treatment of employees, and relations with Third World nations. Moreover, society assumes that blame follows when corporations fail to meet these standards.

We have seen that bureaucratization weakens the accountability of the individual person with respect to corporate actions; thus it appears reasonable to suggest that society's demand must be satisfied by developing *institutional mechanisms* that do not depend on individuals. This solution is analogous to the institutional solutions embodied in liberal democracy and laissez-faire economic theory. Our institutional arrangements in a liberal democracy are designed to prevent individual abuses of power by balancing opposing political interests; and a laissez-faire economic system attempts to obviate the need to rely on individual virtue by automatically converting economic self-interest into public welfare. Whether these mechanisms are thoroughly effective is not the issue; they suggest a strategy for reforming the modern corporation.

But though such mechanisms promise a better fit between corporate behavior and social well-being, few could do much for accountability *per se*. Including public representatives on corporate boards of directors, or instituting "social audit" committees composed of consumers, or giving townspeople "voting shares" of corporate stock—to mention only a few of the current institutional proposals—may limit corporate harm, yet not because of increased *accountability*, but because of a better balancing of competing interests. Protecting people against themselves is not equivalent to making them more accountable. To increase moral accountability there must be an increase in the tendency for individuals and groups to be responsible and account for their behavior in moral terms.

What, then, is the solution to accountability problems in the bureaucratic corporation? Here we must be content with exploring tentative solutions. To block the bureaucratic tendency to destroy individual accountability, we must discover some means to restore individual accountability. Two options appear: first, encourage corporate organizations to return to their historic origins, and reinstitute direct authority relationships between individuals and between corporate strata, thus relieving the need for impersonal rules and reuniting disjoined accountability segments. Second, encourage corporations to push toward a braver, *democratic* model of accountability, in which individuals would participate through institutional structures in the management of the bureaucracy. These alternatives correspond roughly to Dorothy Emmet's distinction between mechanistic and organic organizations, where "mechanistic" refers to those that involve hierarchical systems of control, authority, and communication, with information and knowledge located at the center of the organization, and where "organic" refers to those that have lateral versus vertical directions of communication, continual adjustment and redefinition of individual tasks, and mutual decision-making.[11]

The first model involves a return to classical, Weberian-style organizations, and for this reason seems an unlikely candidate; it is reasonable to assume that the same forces that led to the breakdown of the traditional organization will do the same today. The second is appealing, but sounds dangerously utopian. Can the ideal of participatory democracy survive in the brutal environment of modern business?

Paul Kurtz has argued that the participatory model is the only means of restoring genuine accountability to bureaucratic institutions. "We need," he says, "an organizational bill of rights, an emancipation proclamation by means of which we can build a plurality of democratic institutions."[12] Although not specific about the form such an emancipation might take in the corporation, we may assume that it would involve the participation of corporate employees in decisions that specify corporate goals and define systems of corporate rules (although this participation might be limited). Thus, although features of bureaucratization would remain, e.g., the existence of impersonal rules, employee participation would imply that those generating the rules would be accountable for their nature and impact. Individual accountability for the *application* of impersonal rules would be enhanced, since those applying them would also be responsible for their generation. At the same time, participatory mechanisms would mitigate problems of accountability flowing from centralization and isolation of corporate strata. Participation implies a reversal of the tendency toward centralized, hierarchical decision-making. What is more, *knowledgeable* decision-making implies that participants from one strata of corporate life be acquainted with the facts of life in other strata.

A workable blueprint to specify the parameters of such participation is not presently available—at least not to my knowledge. Nor do we have any reason to suppose that in today's world corporations with participatory structures are likely to emerge. There is evidence that increased worker participation in the design of work conditions yields greater productivity, but the evidence is scanty and must be weighed in terms of the "Hawthorne Effect."[13] (The Hawthorne experiment showed that production tends to increase whenever management alters work conditions, regardless of the change.)

In any case, the ultimate *moral* justification for attempting to solve accountability problems by introducing participatory mechanisms is not greater productivity. It is that our moral ideals require accountability whenever actions are taken affecting the well-being of large numbers of people.

Despite the obvious advantages of the participatory model, there are also costs. Participation would deny employees the luxury of separating themselves spiritually from their work. When one helps *make* the rules, it is harder to say "I just follow the rules." It is harder, in short, to keep one's soul aloof from the organization, and surprisingly, those living in societies that prize individualism are often eager to maintain such an aloofness. Another is that participation from the periphery of an organization does not mean less work for those at the center, it typically means more. When everyone participates in a promotion decision, the pressure to design complex procedures increases.[14] As chairpersons from thoroughly democratic academic departments realize, democracy can be messy. Finally, there is the standard dilemma of all democratic social action; participatory control is only possible through bureaucratic structures, yet bureaucratic structures tend to be destructive of democratic values.

Despite problems, the appeal of participatory mechanisms remains strong. We have seen that the bureaucratization of corporations tends to weaken ordinary accountability. Although fully implementing participatory mechanisms is difficult, it is easy to begin the process. For example, corporations can begin (as some have) to allow a right of dissent or "whistle-blowing" to employees. That is, employees can be given the right to complain about dangerous products or unsafe working conditions *without* suffering penalties. Instituting such a right could be a first step towards enhancing individual accountability.

The final item on the list of accountability problems yields a more obvious solution. Because a bureaucracy increases the power of professionals, and because the new technocratic professions lack altruistic commitments, there is a clear need for programs of ethics at professional schools. Market analysts, chemical engineers, and accountants, as well as doctors and lawyers, must realize that being professionally responsible implies being morally accountable. When technological decisions affect the marketing of a radial tire or the design of a DC-10, then the professionals making those decisions acquire the burden of moral accountability, and they must be intellectually equipped to shoulder it. Here, I think the profession of philosophy can make an important contribution.

Notes and References

[1]These represent three of the four tendencies that Michel Crozier identifies in *The Bureaucratic Phenomenon* (Chicago: University of Chicago Press, 1964).

[2]Crozier, *ibid.*, p. 184.

[3]For Max Weber's view of the paradigmatic organization, see his *The Theory of Social and Economic Organizations* (New York: Macmillan, 1947), see also Dorothy Emmit, *Rules, Roles and Relations* (New York: St. Martin's Press, 1967), pp. 183–215.

[4]For an account of this vicious circle, see Alvin Gouldner, *Patterns of Industrial Bureaucracy* (Glencoe, IL: The Free Press, 1954), especially pp. 172–177.

[5]Paul Kurtz, "The Individual, the Organization, and Participatory Democracy," in *Problems in Contemporary Society*, Paul Kurtz, ed. (Englewood Cliffs, NJ: Prentice-Hall, 1969), p. 193.

[6]John Lachs, " 'I Only Work Here': Mediation and Irresponsibility," in *Ethics, Free Enterprise, and Public Policy*, Richard De George and Joseph Pichler, eds. (New York: Oxford University Press, 1978), pp. 201–213.

[7]Peter Drucker, *Concept of the Corporation* (revised edition) (New York: John Day Co., 1972), p. 88.

[8]Crozier, *op. cit.*, p. 193.

[9]Drucker, *op. cit,* p. 464.

[10]Drucker, *op. int.*, p. 464.

[11]Emmet, *op. cit.*, p. 194.

[12]Kurtz, *op. cit.*, p. 195.

[13]Richard Sennett, "The Boss's New Cloths," *The New York Review of Books* (February 22, 1972), p. 44.

[14]Emmet, *op. cit.*, pp. 195–196.

Evaluating Codes of Professional Ethics

John Kultgen

Introduction

My aim here is to develop a general strategy for evaluating codes of professional ethics. We will examine some provisions of the Code of Ethics of Engineers adopted by the Board of Directors of the Engineers' Council for Professional Development (ECPD) in 1974. This code has been adopted, with modifications and in most cases sans the Guidelines provided by ECPD, by a majority of the engineering societies participating in the Council.[1] We will consider some of the strengths and weaknesses of the ECPD Code in comparison with other codes of professional ethics. We will *not* attempt to reach a final judgment on the merits of the Code because, as I will note, this would require empirical data that are not available at this time. Indeed, one byproduct of the discussion will be an indication of the empirical investigations that are needed to evaluate a code fairly. My critical remarks about the ECPD Code will be designed only to illustrate a strategy, not to carry it out all the way.

Little critical attention has been devoted to professional codes except on the part of those who have devised them, yet the importance of doing so under contemporary conditions of social life is patent. Technical experts are more and more frequently making decisions for the rest of us that vitally affect our well-being. This is especially true of those experts who serve in the professions. Moore notes about the professional in contrast to other skilled specialists such as the artist:

> . . . the welfare of the professional's clients is vitally affected by the competence and quality of the service performed; this is cer-

> tainly not true in the same sense or same degree of esthetic and expressive activities. A poor performance may distress the beholder, but it scarcely threatens his vital interests.[2]

On a larger scale, the professions claim the authority to define values for society in their areas of competence. As Hughes points out,

> Lawyers not only give advice to clients and plead their cases for them; they also develop a philosophy of law—of its nature and its functions, and of the proper way to administer justice. Physicians consider it their prerogative to define the nature of disease and of health, and to determine how medical services ought to be distributed and paid for. Social workers are not content to develop a technique of case work; they concern themselves with social legislation. Every profession considers itself the proper body to set the terms in which some aspect of society, life or nature is to be thought of, and to define the general lines, or even the details, of public policy concerning it.[3]

We may add engineers to the list. They, for example, not only develop sources of energy and the machinery to utilize it, but they shape the public's perception of its energy needs and its policies for meeting them. The professions collectively, then, "presume to tell society what is good and right for it in a broad and crucial aspect of life. Indeed, they set the very terms of thinking about it."[4]

The second feature of modern life that makes it important to evaluate codes of professional ethics is the drive toward professionalism in a broad spectrum of occupations, especially among those that apply advanced branches of science to human problems. "Professionalization" or "the professional project," as it has been called,[5] is the attempt by an occupational group to assume or simulate the organizational characteristics of established professions such as medicine and law and to be recognized as a profession by society. Branches of engineering have been successful in this project, though because of the number of engineers that must be produced to meet the demands of society and the lack of opportunity of most to function as independent agents outside of private and public corporate bodies, they have not been able to acquire quite the social status of physicians and attorneys.

Because of the role that professions typically assume in society once they are established, the spread of professionalism is an important phenomenon. The path that professionalization follows in a particular profession determines how professional life will be conducted, and this is fundamentally important both for the moral integrity of the professional and the welfare of the rest of society. On the individual level, Greenwood points out,

> Professional activity is coming to play a predominant role in the life patterns of increasing numbers of individuals of both sexes, occupying much of their waking moments, providing life goals, determining behavior, and shaping personality.[6]

And we have it on the authority of Talcott Parsons that "many of the most important features of our society are to a considerable extent dependent on the smooth functioning of the professions."[7] In considering the importance of the professions, I emphasize the process of professionalization rather than the properties of a profession in a final and perfected form because the contours of professionalism change with the evolution of society even more in flux in professions on the make such as engineering. A code of professional ethics must be evaluated in terms of the direction in which it propels professionalization in the present state of the particular occupation, rather than in terms of its place in some ideal structure.

This is not to say, however, that ideal conceptions do not play a part in the process that elicits codes. The professional project is governed by a widely accepted ideal-typical model of what a true profession should be.[8] One specification of the model is that a profession is governed by a professional association with a formal code of ethics to which members of the profession subscribe. An occupation striving to be recognized as a profession and following the model will try to develop a code. This is what has happened in engineering in 20th century America.[9]

One's first reaction to codes of professional ethics is that they are A Good Thing—that they cannot hurt and may improve the way important occupations are practiced. But this assumption can be challenged, as we shall see, and in any case some codes are clearly better than others. Both the existence and the content of codes need to be examined critically.

To summarize these thoughts, professions have an important and growing role in society and their codes play a part in professionalization as well as in the subsequent performance of professional tasks. We need to reflect on the criteria for good codes and apply these criteria to actual cases. We need to decide whether code-mongering is a desirable practice and, if it is, how it ought to be conducted and the content that is appropriate for the final product. This paper explores the initial steps in this process.

Understanding Codes

To evaluate a code fairly a critic must, of course, first understand it. A code is a verbal and symbolic entity, a set of formulae whose bearing on professional life depends on the way they are interpreted. The critic must keep in mind that they may be interpreted differently by different groups—in the case of engineering codes, by senior engineers, junior engineers, engineering students, and faculty; by self-employed engineers and those in large organizations; by engineers in different specialties; and by engineers and their clients, employers, regulatory agents, and the general public.

To determine the operational meanings of a code for these groups, the critic cannot rely solely on what its wording means to him or her, or even what it means to its authors or how they intend it to be taken by others. These are sources of interpretative hypotheses about the meanings of the code for its primary subjects, but they are insufficient as a basis for confirming such hypotheses. Definitive judgments about the merits of a code such as ECPD's must await an investigation of the way large numbers of individuals actually respond to it. To my knowledge no one, including ECPD, has committed resources to such an investigation.

This dearth of information does not prevent us from specifying in general terms what interpretive hypotheses about the meanings of a code for its relevant subjects would entail. In doing so we will propose interpretative hypotheses about the ECPD Code. These will be based on my own impressions about engineering practice and the public's image of it. It is my hope that my readers will bring to bear their experience in these matters to confirm or disconfirm the hypotheses, as well as to push forward the dialectic presented here to clarify what such hypotheses entail as propadeutic to evaluating codes.

The most general requirements for interpretative hypotheses about a code are that they should (a) specify its intrinsic or semiotic properties as a verbal entity construed in various ways by relevant groups, and (b) describe its practical consequences insofar as it possesses these properties, in the social context in which it functions. The semiotic dimensions of a code must be determined in order to determine its practical consequences, and its practical consequences are what is assessed in determining the code's value according to the criteria that I will propose.

We will discuss the ECPD Code in terms of its semiotic virtues and deficiences. These are properties that affect the way a code communicates content to its readers within the ordinary framework of vocabulary and syntax of the English language. They include the semantic properties of separate provisions of the code, such as clarity or inclarity, precision or vagueness, and univocity or ambiguity. They include logical properties, such as the dependence, independence, or incompatibility of its provisions, what principles override what, and which adjudicate between others that conflict. And they include the factual and normative presuppositions or the interpretive contexts that induce individuals to construe the code in given ways.

"A code" as a single set of words may be, and usually is, a family of meanings insofar as the words are interpreted in different ways by people with different perspectives. As a consequence, its semiotic properties differ for these groups: A provision may be clear in one interpretation and unclear in another. It may contradict a second provision under one interpretation and not under another. Presuppositions of one interpretation may be true and those of another false.

Other things being equal, clarity, consistency, and truth are desirable in an instrument of communication and in general these properties have good social consequences. They are surely the properties that we would seek in designing perfect codes for a perfect society. In acknowledgment of this, we will refer to them as "semiotic virtues" and their opposites as "semiotic deficiencies." However, virtues are not always good and vices are not always bad in relation to some of one's objectives. A semiotic virtue of a code can be dysfunctional for some of the uses to which it is put and a semiotic deficiency can be functional. A full evaluation of a code must consider the functions and dysfunctions of *both* its virtues and deficiencies. For example, as we will note, clarifying the terms of a code to enhance its value as a guide to conduct for the individual engineer may detract from what we will call its ideological function in promoting the prosperity of the profession.

Semantic Properties of the ECPD Code

Let us now illustrate the kinds of interpretive hypotheses that are required to determine the semiotic properties of the ECPD Code as interpreted by the groups affected by it. The Code begins with four Fundamental Principles (indicated henceforth as P.1, P.2, etc.):

> Engineers uphold and advance the integrity, honor, and dignity of the engineering profession by:

1. Using their knowledge and skill for the enhancement of human welfare.
2. Being honest and impartial, and serving with fidelity the public, their employers, and clients.
3. Striving to increase the competence and prestige of the engineering profession.
4. Supporting the professional and technical societies of their disciplines.

It is inevitable that key terms in the statement of principles in a code designed for a great variety of applications be vague, for example, the glitter words in the ECPD Code, "enhancement of human welfare," "honesty," "impartiality," and "fidelity." They are explicated somewhat, but by no means fully, in seven Fundamental Canons (indicated here as C.1, etc.) and thirty-six sections of seven Guidelines to Practice (G.1.a, etc.). For example, the term "human welfare" is said to cover "the lives, safety, health, and welfare of the public" (C.1). But "the public" is not delimited. It might be restricted to members of the engineer's immediate community (suggested by G.1.e), or it might extend to all living humans, or to all humans including future generations.

The technological mind has been criticized for overemphasizing immediate objectives and neglecting remote consequences and hidden costs, emphasizing calculable factors in a problem and neglecting imponderables, and hence for taking into account obvious and measurable benefits such as health and safety for the immediate community and profits for employers or clients while neglecting more intangible values and the people more indirectly affected by technical decisions. Perhaps in response to this sort of criticism, the authors of the ECPD Code have included an injunction for the engineer to "improve the environment to enhance the quality of life" (G.1.f). No attempt is made to specify how life is to be enhanced, though the reference to the environment may call some sorts of obligations vaguely to mind.

To consider one other attempt to clarify terms, the Canons explicate "honesty" by enjoining engineers to "perform services only in areas of their competence" (C.2), "issue public statements only in an objective and truthful manner" (C.3), and "build their professional reputation on the merit of their service" (C.5). The Guidelines explain something of what these provisions entail, e.g., what C.3 entails for the content of professional reports (G.3.b) and testimony as witnesses (G.3.c), identifying those whom engineers are serving (G.3.d), and the accuracy, dignity, and modesty with which they should comment on their profession (G.3.a) and their own work (G.3.e). The Code thus tries to anticipate the major forms of communication that will be required of the engineer *qua* engineer and specify what honesty entails in each of these. It does not go into detail about the means by which honest communication can be achieved, nor does it deny that other forms of honesty besides honest communication are required. These matters are left to individual judgment.

I do not mean to imply that the ECPD Code does not clarify some matters that were unclear in earlier codes, particularly when it is collated with other engineering codes and interpreted according to opinions expressed by leaders of the profession, such as members of the Ethics Committee of ECPD, the Board of Ethical Review of the National Society of Professional Engineers, and the authors of editorials and articles in the house journals of various engineering societies. Nor am I contending that it would be an easy matter to eliminate residual ambiguities to the satisfaction of the engineering profession and its various publics.

These disclaimers, however, should not blind us to the fact that ambiguities remain that can be and perhaps are exploited for nonmoral purposes. Vagueness and ambiguity have their own uses. Most crucially, the moralistic rhetoric of a code may encourage outsiders to interpret its key terms in their broadest sense to overestimate the commitment of the profession to public service, while allowing the professionals themselves to construe them more narrowly so as to demand less of themselves and their colleagues. One wonders, for example, why the Code goes into considerable detail about the kinds of advertising befitting an engineer (G.5.f, g, h, i, and j), while the very important provision relating to "whistle-blowing" is left ambiguous and vague. It reads,

> Should Engineers have knowledge or reason to believe that another person or firm may be in violation of any of the provisions of these Guidelines, they shall present such information to the proper authority in writing and shall cooperate with the proper authority in furnishing such further information or assistance as may be required. (B.1.d)

A casual reader may assume that this provision imposes on the engineer the duty to "go public" or at least to warn an employer's clients or an appropriate government agency when products or services threaten to be seriously harmful and those ordering the services refuse to do anything about it. The wording, however, allows "proper authority" to be construed narrowly to refer to superiors in the organization for which the engineer works, so that the provision enjoins only whistleblowing on colleagues or low-level superiors, but not on top management or the organization as such. Are engineers to blow the whistle only on other engineers at their own level in the organizational hierarchy, to superiors? On superiors, to top management? On top management, to customers and clients? To government agencies? To the general public through the media and other open forums? Do the profession and its professional associations commit themselves to backing whistle-blowers? To imposing sanctions on engineers who fail to blow the whistle? On employers or regulatory agencies that ignore or penalize whistle-blowers? On these questions, the Code is discreetly silent.

This is consistent with the Code's general approach to matters of global importance. There are only 11 provisions (articles and sub-articles) in the Guidelines for Canon 1 dealing with the obligation to hold paramount the public welfare, compared to 23 for Canon 4 instructing engineers in how to serve as "faithful agents" for employers and clients and 24 for Canon 5 limiting acceptable forms of competition with other engineers. The imbalance suggests that engineers are receptive to specific advice about how to promote the reputation of their profession and prosperity of those who provide their income, but less so about how to serve human welfare. Either they are confident that they can decide this on an individual basis or cannot agree with one another on what welfare entails or believe that the activities covered by the remaining Guidelines automatically take care of the matter.

The Logical Structure of the Code

Vagueness in provisions of a code may be reduced by the context provided by other provisions when the logical structure of the whole is sound and evident. This is not the case with the ECPD Code. As with most codes, it contains no logical asides explaining the relations of the provisions to each other. One is, therefore, left to infer their relations from the sequence of the provisions, their subsectioning, and their content. Thus, one is prone to infer from the content of the Principles of the ECPD Code that provisions listed first in it, and in the Canons and Guidelines that implement each of these, are to be considered more basic than those listed later, at least in the sense that they would take priority over them should conflicts occur (e.g., if a choice has to be made, one should enhance human welfare rather than support a professional society).

An alternative way of viewing the order is to take P.1 (enhancement of human welfare) as a higher-level ordering principle to which one is to appeal to resolve conflicts among any of the others in whatever order the latter occur.

Then again, perhaps the order of the provisions signifies that the justification for later provisions is that they are necessary to implement the earlier ones. For example, the only reason one might be *morally* obligated to support professional societies (P.4), attend their meetings (C.7.c), and so on, and to promote the dignity of the profession by avoiding self-laudatory advertising (G.5.h), abstaining from solicitation of work already assigned to qualified individuals or firms (G.5.d), and so on, is that the reputation of the profession and vigor of its societies can be shown to further the services it renders to the public. Some of these provisions have been criticized as matters of etiquette that are out of place in a moral code.[10] More to the point, they enhance the earning power of professionals and this may compromise the moral authority of the Code. A positive demonstration is required to show that they are morally as well as economically justified. Since the code itself does not engage in argumentation and the presuppositions on which an argument on this point would have to rest are dubious, one must wonder why the provisions were placed in the Code in the first place.

The hypothesis that the order of presentation of provisions represents a logical order among them runs into real difficulties when one turns to the Canons and Guidelines. Neither is listed in order of scope, importance, or instrumental relationships except insofar as each set reflects the order of the Principles. For example, G.1.f, the injunction to improve the environment, is surely more important and comprehensive than those it follows (G.1.a–e). Yet these appear to be a checklist of things to be done in chronological order when dangers to the public arise, with the later ones being necessary only on those rare occasions when the earlier ones fail. The place of concern for the environment at the end of the list thus appears to reduce it to an afterthought required only on rare occasions. Furthermore, its formulation as an open-ended obligation—briefly, with no subprovisions to spell out what it might entail—ignores the possibility that engineering work *per se* entails environmental and other "external" costs in addition to costs ordinarily imposed on the employer or client and that external costs should always be considered in engineering decisions. One may speculate that the authors of the Code felt that a provision about the environment was necessary due to contemporary clamor from environmentalist groups, and perhaps in recognition of the emergence of environmental engineerng as a distinct specialty. Perhaps the provision was included under Canon 1 because the authors did not know where else to put it, not because of any logical connections between it and the other Guidelines for that Canon.

The order of provisions in the Principles, Canons, and Guidelines thus offers little help in determining the logical structure of the Code. Even less help is provided by the order of items within separate provisions. For example, P.1 says that engineers should serve "with fidelity the public, their employers and clients." Most of us would agree that the public should be given priority on those occasions when engineers can determine who "the public" is and how its welfare is affected. Since these matters are often nebulous, however, engineers of conscience are likely to reason that they serve the public best *by* serving their employers and clients. But then trouble is caused by the order of the last two listings. The Code may be read to say that, though some engineers are self-employed and have clients, others work for employers and therefore have no clients of their own, in which case the Code enjoins primary fidelity to the employer in disregard of the well-being of the employer's clients or customers. Another reading of the provision would take the employer's clients to be the clients of the engineer. Then, assuming that the sequence of the listing is significant, fidelity to the employer is said to be more important than fidelity to the client. The provision is morally obtuse on either reading since surely unabridgable loyalty to one's employer is not the best way to serve the public's interest. Thus if one reads the provision to say that it is, one is forced to accept the narrow reading of the whistle-blowing provision (that which does not sanction blowing the whistle on the employer), which raises serious questions about the integrity of the entire code. On the other hand, if the sequence in the listing—public, employers, clients—is not significant at all, the engineer is given no guidance when conflicts of fidelity occur. It is a pious platitude to say that the engineer is obligated to be faithful to all of these groups; it is for the hard cases that he or she needs moral guidance.

This last point illustrates a general limitation of the Code. It may be viewed as a list of *prima facie duties* for the engineer. A *prima facie* duty is one that a moral agent will always consider in determining what is right to do, but one that can be overridden by other duties. A ranking of duties is thus necessary in each case to arrive at a decision. The Code provides no ranking principles or at most vague and incomplete ones. This limits the utility of the Code as a guide to conduct. The Code only lists right-making considerations and leaves it to the conscientious judgment of the individual to determine how to apply them to concrete cases. This, however, may be the best that any code can do to define the obligations of thousands of agents in a multitude of diverse circumstances. Any attempt to go further in the direction of ranking duties according to general principles might cause more harm than good. I am not convinced that this is so, but I am aware that it is not possible or desirable to eliminate personal judgment entirely by an ethical code.

Presuppositions of the ECPD Code: For the Uninformed Public

The presuppositions of a document comprise the factual and normative beliefs of the major groups who read it and determine the sense in which they take its words. Semantic and logical inclarities of a code open the door for varying interpretations of its provisions according to one's presuppositions, which, we have suggested, will differ among the groups responding to the code. We will consider first some presuppositions that are likely to be entertained by those relatively ignorant of engineering practice when they read the ECPD Code (as I myself was when I first became aware of it). Then we will note the falsity of these presuppositions and investigate the effect of a more adequate understanding of engineering practice on the way the Code is likely to be taken. The reader should keep in mind that our analysis is hypothetical since it bears on the attidudes of actual groups of people, and that our evidence for the hypothesis is impressionistic and limited.

To one unfamiliar with engineering, the Code encourages the misconception that engineering is a single profession. The Code emanates from a consortium of major engineering societies. Its title proclaims the common purpose of "professional development" (development of engineering as a profession, or of engineers as professionals, or of professionalism among engineers?). The Council has published the definition:

> Engineering is the profession in which a knowledge of the mathematical and natural sciences gained by study, experience, and practice is applied with judgment to develop ways to utilize, economically, the materials and forces of nature for the benefit of mankind.[11]

It thus depicts engineering as a single whole. And the Code itself undertakes to specify the obligations of "Engineers" without distinction as to specialty or employment.

I do not deny that engineering can be considered a single profession in some sense of that nebulous term, but I wish to call attention to another sense in which engineering is clearly not a single profession and to the effect of confusing these senses on the interpretation of the ECPD Code.

I mentioned above a widely accepted model of professions in sociology that is based on paradigm cases in the history of professionalization—namely, medicine, law, the ministry, and to some extent architecture and the military.[12] To present one version of the model succinctly: A profession is an occupation that (a) applies a single body of theory, over which the occupational group has a practical monopoly, to problems of vital importance to others (clients, patients, customers, sectors of the public), an activity that (b) requires advanced education, usually in a university setting, and life-long practice and development, and (c) is primarily oriented toward the good of society rather than profit, (d) whose performance is organized and regulated by the group itself, through an association that typically assumes certain characteristics, among which is (e) the adoption of a formal code of ethics and the inculcation of its principles in the professional conscience of the practitioner.

When this model is utilized by a group striving to professionalize itself or to justify its claim to be a profession already, it is usually not made clear whether the model is to be taken as an unrealized ideal, a typification of paradigm professions, or an empirical generalization that applies to all recognized professions.

Whatever the intent of groups that utilize the model, it does not in fact accurately represent the features of even paradigm professions without extensive qualifications for each of the properties we have listed. It is even less accurate as a representation of more recent entrants into the ranks of the professions, especially since emerging occupations may falsely simulate the properties now that the sociological conception has gained currency and can inform their public relations efforts. As one observer has noted:

> Particular tactics adopted by one or another occupational group, organized by an association, recognize *criteria* of professionalism, if not a scale or sequence. The result may well be a kind of checklist approach, once the criteria are enumerated with some degree of common assent. Since symbols may hide the absence of reality, and the manipulation of symbols is often easier than the changing of actual organizations and bahavior patterns, it is scarcely surprising that a considerable part of the observed behavior of organized technical occupations consist in the conscientious manipulation of symbols. One can almost envision the process of self-authentication by an occupational group: Service orientation? Provided in our adopted

> code. High admissions standards? We have achieved a university curriculum. Autonomy? Well, in our salaried position we do the best we can.[13]

The general thesis that the model of professions is misleading need not be argued here.[14] What is to the point are the ways in which engineering fails to conform to the model and hence is not a profession in the way the model encourages observers to think, despite the fact that ''real'' engineers are professionals in a legitimate sense of the term.

The Facts of the Case

The reason why the lack of unity in engineering is significant in this context is to be found in the concept of autonomy in the model of professions. An occupational group claims special rights and privileges in society by virtue of being a profession. Perhaps the most basic is monopoly of expertise in some important area of human life. For example, medical doctors claim an exclusive right to manage serious disease and enlist society, especially agencies of government, to limit rival services by other groups such as chiropractors, naturopaths, and midwives, and even their own auxiliaries such as nurses, druggists, psychologists, and therapists. To justify these measures medicine guarantees not only a level of competence, but an ethical standard of practice by its own members. To foster these it controls the education and licensure of practitioners and monitors their work. A code of ethics is adopted both to inform the practitioner what is expected of him or her and to proclaim to the public that the group is worthy of the trust reposed in it by acquiescence in its monopoly. The justification for a profession's monopoly of expertise, thus, is its ability to govern itself effectively, as well as its disposition to do so for the benefit of society.

Can engineers claim to be an integrated self-governing group so structured as to be safely entrusted with the exclusive right, in major and complex matters, to ''apply knowledge of the mathematical and natural sciences to utilize the materials and forces of nature for the benefit of mankind''? Engineering is sufficiently *not* an integrated profession in two respects to give us pause: (a) professional engineers are dispersed among a number of professional and quasiprofessional groups according to the particular branches of science that they employ, and (b) not all ''engineers'' are professionals in the sense implied by the model.

A profession, in the strict sense of the term, must be dominated by a single professional association. Autonomy entails self-regulation by an organization capable of expressing the collective will of the occupational group and taking responsibility for the way individuals in it do their work. The reader will find some thirty-one associations for engineers listed in Appendix I (together with their acronyms). They claim a combined membership approaching 600,000. Only two, ECPD and NSPE, represent engineers across the engineering specialities. ECPD is a council of other associations, including most of those of the most established specialties, but these include only about half of all engineers in associations. NSPE is designed for engineers registered by state boards of examiners and has a membership of only 68,000.[15] The ECPD and NSPE Codes of Ethics are, therefore, largely advisory to specialized associations, which, in respect to most organized engineers, possess the measures of enforcement. For the most part, the specialized associations have either failed to adopt the general codes or have modified them extensively. We should note in particular that a number have adopted the Principles and Canons of the ECPD Code without the Guidelines, reducing it to a set of pious platitudes. Other associations have developed parallel codes, in some cases fairly detailed, in others, general and bland. SAE notes haughtily:

> While fully subscribing to ethical guides, SAE has no formally enunciated code of ethics. The members have generally felt that they know what ethical behavior is and have lived up to it as well as any other engineering group; they refuse to adopt a code merely to be numbered among Societies having such a Code.[16]

The dispersal of engineers among subgroups and the multiplicity of their codes does not, of course, necessarily mean that engineers do not possess a common ethic. The various codes share a number of themes at the level of principles and sometimes in matters of detail, and many variations in detail can be traced to differences in the recurrent ethical situations in which, for example, consulting engineers and fire protection engineers, find themselves. The variations in the codes, however, tend to blur the responsibilities of the engineer as such and partly account for the vagueness of the ECPD Code as it tries to be adaptable to the needs of various groups.

To the extent that engineering is ruled by formal organizations, then, it is a cluster of professions rather than an integrated whole. One cannot look to any one document, such as the ECPD Code, for a definitive statement of its principles of action. Furthermore, the definition of "engineer" on the individual level is much more uncertain than, say, "physician" or "attorney." A primary criterion for membership in some of the major specialized associations is the completion of a course of study approved by ECPD (e.g., as AIChE, AIIE, ASAE, ASCE, and NICE) or other "approved program" (AACE, AIPE, ANS, ASME, IEEE, SAE, SAME, SPE), many cases with the additional requirement of employment in the field for a specified length of time. Sometimes work experience is allowed to substitute for formal education, and other combinations of credentials are allowed especially in associations with hybrid memberships (e.g., AAEE, ASHRAE, ASSE, SFPE, and SPHE). Besides Members, some associations admit Fellows, Honorary and Senior Members, who may have stronger qualifications, and Associate and Affiliate Members, with weaker ones. All of these with some right may call themselves engineers. In addition, there are uncounted numbers of individuals with engineering degrees who do not belong to associations and technicians doing engineering work on the basis of noncertified educational programs and on-the-job training. Indeed, engineering graduates complain of being hired as engineers and then assigned the work of "mere technicians." It is clear that engineers as a group have not succeeded in drawing a sharp line between their membership and individuals offering competitive services who are not under the discipline of engineering associations. Even those who are clearly entitled to the name "engineer" are so diversified as to make problematic the assumption that they have a common ethic or that the ECPD Code speaks for the whole when it specifies what "Engineers" do, e.g., what engineers recognize they ought to do when faced with ethical choices on the job.

The Code from Above and Below

Engineering is not *a* profession with the kind of mechanisms for regulation of its members that is the basis for the claim of autonomy of the paradigm professions. This fact casts light on the peculiar imbalance in the provisions of the ECPD Code and its irrelevance to the ethical dilemmas that confront the majority of practicing engineers, most of whom occupy subordinate positions in commercial and governmental organizations. Rothstein notes,

> Engineers work in many different situations in which they are subject to group and organizational control. For example, engineers are usually members of primary work groups that have norms and values developed by the groups, which may be composed of engineers and non-engineers. If engineers, as primary group members, deviate from those norms and values, they will be punished by their fellow group members. Engineers are also members of formal or-

> ganizations and are subject to the organization's rules, which are probably not devised by engineers. If they violate the rules, they are likely to be punished by their organizational superiors, who may not be engineers. Because engineers work in thousands of different work groups and organizations, they are subject to thousands of sets of norms, values, and rules, many of which are undoubtedly dissimilar and even conflicting.[17]

Rothstein cites this fact to deny that engineering is a self-regulating profession. But it also means that, if there *are* to be operative norms specific to engineering and if, as is inevitable, these conflict with demands of groups in which engineers work, those who are faced with such demands must be provided with guides tailored to their situation and incentives to do what professionalism requires despite the sacrifice that this involves.

It has been frequently noted[18] that engineering codes were originally modeled after the codes of the paradigm professions, especially medicine and the law. At the time *they* were devised, professionals typically were independent agents who entered into voluntary fiduciary relationships with individual patients or clients. Engineering codes as a consequence have been oriented toward self-employed consultants and engineers in higher level management positions who make decisions for the organizations in which they work.

Thus, most of the Guidelines for implementation of Canons 3, 4, and 5 of the ECPD Code (relating to public statements, conflicts of interest, and competition with other engineers) are appropriate to the ethical questions facing this rather small group and have little application to the situation of engineers in subordinate positions in large organizations. Those provisions that do apply specifically to subordinates, for example, some of the Guidelines for Canon 4, seem designed to assure employers that engineers will be faithful employees, in the hope that they will treat them like professionals rather than skilled workers.

Perhaps the reason codes are written with the independent agent in mind is that their authors generally belong to this category themselves. Leaders of professional associations are those who can afford to participate in national meetings and are in a position to gain release time from their work to do so. They are also individuals who enjoy the prestige to win leadership posts owing to their financial success through engineering accomplishments or through entry into management structures.

One critical bit of evidence that this is the case is the attitude of leaders of the engineering professions toward unionization. It appears to be considered not just inexpedient, but faintly immoral for engineers to join unions. The NSPE Code does not prohibit unionization, but it equates strikes with coercion and condemns both:

> Section 1—The Engineer will be guided in all his professional relations by the highest standards of integrity, and will act in professional matters for each client or employer as a faithful agent or trustee. . . . .
>
> f. He will not actively participate in strikes, picket lines, or other coercive action.

The ECPD Code does not contain a specific provision to this effect, but ECPD has endorsed the Guidelines to Professional Employment for Engineers and Scientists, which enjoins employers not to accept closed shops that include engineers:

> Terms of Employment:
>
> 13. There should be no employer policy which requires a professional employee to join a labor organization as a condition of continued employment.

I think that the implication is clear that the subordinate engineers are not to engage in collective action that might give them real influence over their work conditions. They are to leave it to employers, as persuaded by leaders of the profession, to treat them individually as befits professionals.

As for the ECPD Code, the inapplicability of Canons 3, 4, and 5 leaves the subordinate engineer with Canons 1, 2, 6, and 7, insofar as they are generally applicable to all engineers. These relate, respectively, to promoting the public interest, practicing only in one's area of competence, associating only with reputable persons and organizations, and continuation of professional development. In the absence of guidelines with bite, acceptance of the Code is on a par with taking a stand for God, Motherhood, and Apple Pie. For the subordinate engineer, it is essentially a ceremonial act.

To summarize the suggestions in this discussion of presuppositions, the vagueness of a code is a function of the diversity of those to whom it is addressed: vagueness is necessary to assure a favorable hearing among these groups. Vagueness allows the code to be interpreted differently according to the presuppositions that the groups bring to it. We have distinguished in particular between the presuppositions about the engineering profession of people largely unfamiliar with its structure and those familiar with it.

Those unfamiliar with engineering are likely to assume that it is a single profession in the strict sense and look upon a code ostensibly designed for all engineers and containing detailed provisions, such as the ECPD Code, as an effective and binding document that guarantees the public that engineers will do their work in an ethical manner. Leaders of the profession encourage this impression by exhortations to engineers to be faithful to the ideal of public service.

Among those familiar with the engineering profession, a small group, whose attitudes are shared by leaders who draw up codes, may find the ECPD Code helpful when they wrestle with ethical problems.[19] The response of this group perhaps encourages the authors of the Code to think they are speaking to and for the profession in promulgating it.

The large part of the engineering profession find that the Code is largely irrelevant to the dilemmas they face in the workplace. They are left to their personal code, obtained elsewhere than from the profession, to determine whether they are obligated to accede or resist demands of their work group and organizational superiors. For them the Code is largely ceremonial except to the extent that it encourages faithful obedience to the employer.

Codes of the profession thus have the effect of consolidating a division of engineers into a fairly cohesive elite and the scattered remainder. The latter are encouraged to let employers, under the guidance of the leaders of the profession, make moral decisions for them to the extent that these are involved in assigning them work and disposing of its products. Moral paternalism on the part of the leaders is strongly implied, since it is they who try to determine for the whole profession the moral orientation that it urges upon employers and the public.

Evaluating the Consequences of a Code

We began our discussion with the truism that a critic must understand a code in order to evaluate it. We claimed that understanding involves a determination of what its words mean to those aware of it and the practical consequences of their interpretations. We found that determining a code's meaning is not easy since those affected interpret it in various and complex ways. Nor can the question of interpretation be entirely divorced from the question of consequences. The reader will have noticed that we have pointed to some consequences as evidence that the ECPD Code *is* interpreted in certain ways and, to the extent that its authors intend those consequences, we used them to explain features of the Code. It is now time to deal with consequences systematically.

We will begin by discussing the evaluative criteria that we will have in mind in selecting consequences for description. Up to this point, we have attempted to present the semiotic features of the ECPD Code as neutrally as possible, but obviously we were interested in such matters as vagueness and clarity, relevance and irrelevance, and so on, because we anticipated that these affect the good and harm that the Code does for the profession and society.

The ethical perspective from which we will view the consequences of the ECPD Code is a form of Ideal Rule Utilitarianism adapted from Richard Brandt.[20] I shall modify and extend Brandt's theory *ad libitum* to deal with the topic at hand.

The premises of Ideal Rule Utilitarianism pertinent to our inquiry are these:

> (1) The utility of an action, practice, or rule is its contribution to the happiness or reduction of the unhappiness, less the unhappiness it causes and happiness it reduces, of the sentient beings affected by it.
>
> (2) The conditions of happiness are variable from individual to individual and only approximately predictable. (Hence, actual actions and rules must be evaluated in terms of their probable utility.)
>
> (3) However, for almost every human being, happiness entails moral activity, both because of its intrinsic satisfactions, the sense of belonging to a moral community that it brings, and its tendency to elicit moral actions from others.
>
> (4) The moral rightness of an act is determined by the utility of the moral rule to which it conforms or the utility of the performance of all or most of the acts of the class to which it belongs.
>
> (5) The rules that determine the morality of an act fit into an ideal moral code. An ideal moral code is that among viable alternatives which, were it to gain currency within the framework of the present institutions of society, would maximize utility among those subject to the code and affected by their actions.
>
> (6) The ideal moral code is the basis for judging the morality of the practices of an institution and striving to change them. (Once they are changed, the ideal moral code may have to be revised to be ideal for the new institutional framework.)
>
> (7) An ideal institutional code is that which, were it to gain currency within the framework of the moral code and the other institutions of society, would maximize utility.[21]

To apply these premises to engineering codes, we treat engineering as an institution, that is, as a set of coordinate roles, each with norms reflected in the expectations of those involved in or dealing with the institution and enforced by formal or informal sanctions. The norms determine interrelationships among role occupants and define their rights and obligations *vis-a-vis* one another. In the case of professionals, norms typically define their relations, rights, and obligations toward clients and employers, colleagues and "the profession," and bystanders and "the public." These rights and obligations may be spelled out in terms of the individual's relation to institutions, e.g., the firm, the professional association, the government, and so on.

The norms of professional ethics may be distinguished analytically from two other sorts of norms, though they are often intermixed with them in practice: (a) Technical norms of competence define what the work of the profession is and how it must be done to be done well. (b) Prudential norms bear on the way the professional and the profession conduct themselves so as to win or preserve a socioeconomic position. (c) Ethical norms bear on whether work ought to be done, to what end, with what safeguards, under what conditions, and so on.

All of the norms are to be judged ultimately in terms of their utility. Ethical norms are the focus of our interest here. The first thing to note about them is that we are speaking of the operative ethic of a profession—the norms that are actually observed by most practitioners and used by them to judge one another's professional actions as ethical or unethical. These may be stated in a formal code and this may be what authors of a code try to do. But there also may be a divergence between operative and formal norms and authors may try to alter the former by promulgating the latter. One who is evaluating a formal code must first determine its relation to practice. Does it reflect or promote operative norms and do these maximize the utility which the profession is capable of generating? And does the code have other utilities or disutilities?

Once again it is useful to distinguish between understanding and evaluating a code, particularly since the term "functions" has a dual use. The *social functions* of a code are its contributions to the development or perpetuation of systems of action, in this case, the action system that constitutes the work of the profession and the larger system that constitutes society and all of its institutions, whose demands the profession is designed to meet. Success is defined in terms of the desires of the professionals and their patrons for a certain formation of that occupation. To the extent that a code promotes the system of action that they favor, it is socially functional. To the extent that it does not, it is socially dysfunctional. *Sociological understanding* of a code consists in a determination of its social functions and dysfunctions.

A sociological understanding of a code is necessary, but not sufficient, for a *philosophical evaluation* of it. It is necessary because social functions and dysfunctions are primary features of a code that need to be evaluated. It is not sufficient because action systems can endure for long periods of time and nevertheless fail to maximize utility. One must determine whether the social functions of a code are *rational functions* as measured by the principle of utility. Do its principles express or promote operative norms that insure that the profession will make its optimal contribution to human welfare within the framework of existing morality and other institutions? Does the code have other utilities and disutilities? A code of ethics that shapes an occupation according to the desires of its authors and their patrons may be rationally dysfunctional when judged by this criterion.[22]

We shall distinguish two basic sorts of social functions, regulative and ideological. Under regulative functions, we will consider guidance, judicial, and contractural uses of ethical codes. Under ideological functions, we will consider ceremonial, rationalizational, and public-relational uses.

Regulative Functions of Codes: Guidance

By the regulative functions we will mean the use of a code to shape the conduct of practitioners of the profession *qua* professionals in the typical situations in which they find themselves in their work, especially in those in which there is a conflict between morality and self-interest or among moral obligations. The leaders of a profession or the professional association might leave such decisions up to the individual or they might attempt to promote adherence to common norms. Codes of professional ethics are ostensibly instruments for the latter.

The first question to be addressed is whether the aim is a desirable one. *If* the engineering profession could agree on a code that would bear critical scrutiny under the criteria of Rule Utilitarianism and *if* the code could gain currency in practice among engineers, the aim would seem to be a good one; but even this needs to be argued. What Brandt says about moral rules in general applies to special rules for professionals:

> We must remember that it is a serious matter to have a moral rule at all, for moral rules take conduct out of the realm of preference and free decision. So, for the recognition of a certain moral rule to have good consequences, the benefits of recognition must outweigh the costliness of restricting freedom.[23]

Why impose on engineers the burden of professional obligations in addition to those of ordinary morality and the demands of their employers and work groups? The answer is fairly complex, and bears on the limitations of the ECPD Code.

The first consideration pertains to the guidance use of a code. Its premise is that engineers face special ethical questions for which routine moral training does not adequately prepare them (a claim that only an extensive study of actual cases can document). Now, a good professional code according to Rule Utilitarianism must be consistent for the most part with ordinary morality and applies the general principles of the moral code current in society to the special circumstances of the profession. However, ordinary morality is not without its internal contradictions and dubious elements, and its implications for special circumstances are not always obvious. For example, ordinary morality enjoins us to tell the truth, but not necessarily the whole truth or nothing but the truth in all circumstances. The ECPD Code spells out in Guideline 4 some of the truths that must and must not be told in engineering:

> a. Engineers shall avoid all known conflicts of interest with their employers or clients and shall promptly inform their employers or clients of any business association, interests, or circumstances which could influence their judgment or the quality of their services.
>
> h. When, as a result of their studies, Engineers believe a project will not be successful, they shall so advise their employer or client.
>
> i. Engineers shall treat information coming to them in the course of their assignments as confidential, and shall not use such information as a means of making personal profit if such action is adverse to the interests of their clients, their employers, or the public.
>
> n. Engineers shall not attempt to attract an employee from another employer by false or misleading represenations.

Some of these provisions appear to be sanctifications of the obvious, yet it may be that morally naive engineers need to have the temptations and obligations in this area spelled out. Perhaps a stimulus is useful to induce them to reflect on the more subtle problems of truth-telling in real situations.

But why foster specifically *moral* solutions to the special problems of engineers? The obvious answer from the standpoint of the Utilitarian is the importance of engineering decisions for the rest of humanity. Society has a stake in seeing that the action system that is the profession of engineering be directed toward maximum social utility. That is what its operative ethical norms are for. I should like to emphasize in addition the proposition (Number 3 above) that moral activity is essential to the happiness of the individual engineer. The factual question here is whether guidance and encouragement from fellow engineers contributes to this end given the confusions and pressures of the workplace.

I have my doubts about guidance in the paternalistic form in which it is sometimes offered. The author of a manual on professionalism suggests, "A young man leaving college is plunged frequently into a life so different from his previous experience that he needs advice of the men who have already had the experience that is about to become his."[24] He expresses his desire to preserve such men from corruption by imparting the lessons of fifty-six years of professional practice in the "school of experience and 'hard knocks.'" Setting aside the author's apparent inability to conceive that women might have a place in engineering, I would point out that he seems to think that ethical principles can be imprinted at a formative period of life and do not ever need to be reconsidered. Ethical counsel is therefore not needed by mature and successful practitioners. One does not find such complacency among those who have grappled with real ethical questions on either the theoretical or practical plane. The engineering profession should be aware that a serious expedition into the bramble fields of ethical reflection will raise more questions than it settles. It is my sense, therefore, that the guidance value of an effort by engineers to reach consensus on an ethical code, and to review and revise it periodically to meet changing conditions, will be as useful as the final product itself. The value of the enterprise lies in arousing engineers, mature and established as well as novice, to reflection about the ethical implications of their work, rather than in providing stagnant recipes for action.

Regulative Functions of Codes: Contracts

The value of drawing all engineers into the forum of ethical dialog makes the elitist organization of engineering associations and the irrelevance of their codes to the problems of the dispersed majority a critical problem. There is a tendency in any large organization to reduce subordinates to links in a causal chain, rather than autonomous moral agents. They are encouraged to view their work as the exercise of technical competence to carry out small steps in a collective effort whose direction is determined from above. The good engineer in such an environment is amoral like the good computer. Vivian Weil has observed,[25]

> Since over 75% of all engineers are employed in large organizations, many of them are vulnerable to . . . (a) sort of fragmentation, the separation of moral obligations of private life from the obligations governing professional activity . . . the schizophrenic isolation of one's moral self from one's professional self.

The moral which she draws is as follows:

> What reasons can be given for allowing human conduct in economic, governmental, and other organizations to be so reduced? Why should we think that the ethical problems which arise for individuals in those settings give rise to special privileges that bypass ordinary moral obligations? One answer might be that morally sensitive behavior demands too much personal sacrifice. But that points to the need for change in the work setting, for a search for feasible reforms.

The specific reform which she explores is that in the engineer's self-image. A way needs to be found to enable engineers to view themselves as moral agents, not as mere causal factors in a super-personal mechanism. I have suggested that one step in this direction would be the development of a code of engineering ethics for the bureaucratic setting to supplement the provisions for the enterpreneural setting found in the ECPD Code. But exhortations to truthfulness and the like ring hollow to the galley slave; guidelines to action are feckless for those who lack the power to act. An effective ethical code would require structural changes in the organizations in which the work is done. Unger argues,

> . . . a fundamental step would be to transform the work environment of the engineer to enable him to function as a professional rather than as a cog in a machine.
>
> In such an environment, the engineer must be free to discuss work goals with management, to exercise much greater discretion in implementing these goals, and to decline assignments on ethical grounds. Where his professional judgment is overruled in serious matters involving the public welfare, and where protests to his management are of no avail, the engineer should be free to take his case to professional societies, governmental bodies, or any other appropriate forum.
>
> Provided that he acts within the framework of the code of ethics, interpreted in the light of his own values, he should be defended by his professional society against retaliation by his management. This

> should be the case regardless of whether he is employed by a governmental agency, a commercial firm, or any other type of institution.[26]

Unger suggests a number of measures to promote the treatment of engineers as professionals, following the model advocated by the American Association of University Professors for academia. They include a tenure system with safeguards against arbitrary dismissal, portable pension plans, and sanctions by professional societies against censured organizations.

The effect of these on ethical conduct would be slight apart from effective guarantees of the employee's rights of conscience by the organization.[27] The relevance of this to codes of ethics is that they can usefully contain a statement of rights of the engineer, i.e., of ethical claims for specific kinds of treatment by private and public sector employers, regulatory agencies, professional associations, and other institutions. Without the power to act there are no obligations, and obligations, conversely, require the right to act. Thus a draft code for the American Academy for the Advancement of Science recently discussed by Robert Baum[28] devotes sections not only to the responsibilities of scientific workers, but to the rights necessary to carry out those responsibilities and the obligations of professional associations "to provide support for members complying with the Code as well as to penalize members who violate the Code."

It is here that what I call the contractual use of codes could come into play. Acceptance of a code containing rights as well as obligations might be viewed as a contract between the individual and his or her colleagues. The individual promises to use the skill acquired with the help of the profession and exercised in situations of opportunity opened up by its guarantees of competence in an ethical manner, as defined by the profession's code. The profession promises to back the individual should he or she get into difficulty through adhering to the code.

More important perhaps would be contractual relations between the individual and the employer. The Guidelines to Professional Employment for Engineers and Scientists[29] has won greater acceptance in the engineering profession than any other policy document. It is supported by ECPD and NSPE as well as twenty-five other associations, including those of the major specialties in engineering. It is proposed to engineers and employers alike as a statement of appropriate conditions of employment. Although the Guidelines call on employers to foster "ethical practices" and recognize "the responsibility of the professional employee to safeguard the public interest" (Objectives 1 and

3), they make little attempt to define ethical practice or the requirements for safeguarding the public interest. They make no reference to the profession's code of ethics. They do contain cognate provisions, but these are ones that deal primarily with the employee's obligations to the employer, e.g., relating to loyalty, confidentiality, and diligence, and the employer's obligations to the employee, to treat him or her as a professional, not to use a time clock, vesting of pension rights, and so on. These provisions are designed to insure a fair exchange, to gain a position for the engineer superior to that of the skilled worker, and to provide the employer greater loyalty and creativity in return. They have little to do with service to human welfare by either employee or employer, except to the extent we can assume that successful employers are a boon to society. The one reference to public welfare in the Guidelines is as general and nebulous as Canon 1 of the ECPD Code:

> The professional employee should have due regard for the safety, life, and health of the public and fellow employees in all work for which he is responsible. (Term of Employment 2)

This single provision is surrounded by some forty-five major requirements relating to material conditions of recruitment, employment, professional development, and termination, similar to the sort of things that are of primary concern to labor unions.

The unrealized possibility that I have in mind here is an agreement throughout the engineering profession to refuse employment unless the individual's right to follow the profession's code of ethics is guaranteed. This would make sense, of course, only if the profession could agree upon a code or a set of parallel codes that met the specifications we have been discussing, including an adequate statement of employee rights.

Regulative Functions of Codes: Judicial Proceedings

We have noted that a formal code of ethics has an effect on the practice of a profession to the extent that it shapes the operative norms of individuals directly by giving them guidance as to what is ethically required, and that it could be indirectly effective if the right and obligation to follow the code entered contractural relationships. Each of these uses may be supplemented by the use of the code in what we will call judicial proceedings. To see why these might be needed in professions, we must reflect further on the conditions for effectiveness of ethical norms.

For a code to be effective, it needs not only to be sufficiently clear, specific, and morally defensible to guide a morally disposed individual, and relevant to situations in the workplace where he or she has the power to make decisions, but it must be enforced by sufficient sanctions. Brandt points out that according to Ideal Rule Utilitarianism codes are ideal according to whether they maximize utility in conditions that actually obtain in a society. For most societies this means that they must be binding for persons of ordinary conscientiousness as well as saints.[30] I take this to mean that an ideal code does not require heroic virtue or extreme sacrifices on the part of most people to do what is moral under ordinary circumstances. This is why the content of a code of engineering ethics must be designed with an eye to the rights that engineers enjoy, or might win, in the workplaces of society as it is now organized. A code must also appeal to what Baier calls "people of limited goodwill." People of absolute goodwill do what is right come hell or high water. People of limited goodwill are "those who acknowledge the legitimacy of moral rules and are prepared to obey them, even in the absence of deterring sanctions, but who are prepared to do so only on condition that others are doing likewise."[31] A code needs sanctions not only to deter malfeasants, but to assure persons of limited goodwill that malfeasants will be deterred. Another way to put this is to say that it is meaningless, at least by utilitarian standards, to make people responsible *for* certain actions unless they can be made responsible *to,* i.e., answerable to, someone with the power to inflict unwanted treatment on them.[32]

In a classic statement of the connection between morality and punishment, Mill makes clear the primary kinds of sanctions that are available and who exercises them:

> We do not call anything wrong unless we mean to imply that a person ought to be punished in some way or other for doing it—if not by law, by the opinion of his fellow creatures; if not by opinion, by the reproaches of his own conscience. This seems the real turning point of the distinction between morality and simple expediency.[33]

Most professions rely on the internal sanction of professional conscience, inculcated in practitioners during their socialization into the professional subculture, and the informal external sanction of peer criticism, incurred during the career of professional practice, to insure that most professionals conform to the operative norms of the profession. A few professional associations, notably, medical associations and the bar, have been able to enlist the formal sanctions of the law. This requires a system of licensure as a condition of practice, designed by the profession to exclude nonmembers and enforced by the state, and a judicial–punitive apparatus to detect and punish violators. In such a system, the profession's code of ethics takes on a quasilegal status, and this is what we mean by the judicial use of the code.

Short of penetration into the formal institutions of the state, a professional association has only the power to reprimand members or expel them from its ranks. This is also a judicial use of codes, but it has no effect on nonmembers and a limited effect on unethical members, since membership in a professional association is not necessary to make a good living where there is no licensure.

Some of the leaders of the engineering profession are clearly envious of professions that are more entrenched in the political system, but it needs to be demonstrated that formal sanctions would significantly raise the ethical level of engineering practice. We have suggested that the first condition that would have to be met would be the perfection of a code that is acceptable on moral grounds and relevant to the work conditions of the whole profession. A second would be important structural changes in employing organizations. Standing in the way of both is the size and heterogeneity of the profession and the maladaption of engineering associations for the reformative functions we have imagined. It may be that voluntary adherence to codes is the most that can be expected under such conditions.

Indeed, it can be argued that voluntary use of a code by individuals is the only truly moral role that one can have. It may be thought that judicial uses of codes, accompanied by state sanctions or even the weak sanctions of the professional association, fatally compromise them as ethical documents. Thus, John Ladd maintains,[34]

> Ethics is an open-ended, reflective and critical intellectual activity. It is essentially controversial, both as far as its principles are concerned and in its application. These principles are not the kind of thing that can be settled by fiat, by agreement or by authority (e.g., by professional associations).

On the basis of this concept of ethics, he concludes,

> The attempt, e.g., by professional associations, to *impose* principles on others in the guise of ethics contradicts the notion of ethics itself, which presumes that persons are autonomous moral agents. Attaching disciplinary procedures and sanctions to principles that one calls "ethical" automatically converts them into legal rules or quasi-legal rules, conventions or regulations.

Ladd utilizes a defensible, but not by any means the only legitimate sense of the term "ethics." The sense with which we have been operating is closer to what professionals have in mind when *they* talk about "professional ethics." Whatever group standards be called, they are an important phenomenon and need to be evaluated by principles of utility, both in their content and in the way they gain currency. Ladd's remarks call attention to the importance of the latter. I have agreed with him (so far as his position can be inferred from the brief passages above) that an individual is morally bound by a code only if he or she has had an appropriate role in forming or reviewing it and can see that it is appropriate, at least in the main, for the professional context. We have noted that consensus is very difficult to achieve in engineering's fragmented state. Nevertheless, in striving to nurture ethical practice, the options are not just for some to impose principles on others or to abandon the hope for common principles altogether. They include the attempt to develop a consensus and have it "enforced" primarily by the informed conscience of persons of goodwill. In this, a limited role may be played by judicial and contractual uses of a code to protect the opportunity of persons of good will to follow it voluntarily, but Ladd's remarks do remind us that these measures should be used with utmost caution.

We asked earlier whether the engineering profession should attempt to use ethical codes to regulate professional practice, if an adequate code could be developed and gain currency among engineers. I will reiterate the judgment that it should try; but now we must add that it must proceed cautiously and that prospects of currency are not very bright. And of course, the development of an adequate code is still a *desideratum*.

Does this mean that the ECPD Code has no utility or only the limited utility of its modest regulative function? Not necessarily. We must turn next to the ideological functions of codes, which also demand philosophical evaluation in terms of principles of utility.

The Ideological Function

Goode notes that to be a profession, an occupation must not only possess certain properties, it must be perceived by society to possess them. He identifies two "generative traits," from which the remainder of ideal-typical traits of a profession are derived, (a) monopoly over a body of abstract knowledge applicable to concrete problems of living, and (b) the ideal of service as the "collectivity orientation" of the group; and he asserts:

> The society or its relevant members should believe that the knowledge can actually solve these problems (it is not necessary that the knowledge actually solve them, only that people believe in its capacity to solve them). Members of the society should also accept as proper that these problems be given over to some occupational group for solution . . . because the occupational group possesses that knowledge and others do not. . . . The society actually believes that the profession not only accepts these ideals (of service) but also follows them to some extent.[35]

A professional ideology is a complex of ideas that an occupational group utilizes to achieve or present its status as a profession.[36] The group adapts the sociological model of professions discussed above to its own conditions so as to present itself as a true profession.

Why are professional ideologies needed in the professional project, and what are the ideological functions of codes of professional ethics? Professions need an ideology because they claim special rights and privileges for which society must pay a cost. To enjoy many of these rights and privileges, other groups must be denied them. Professionalization thus becomes a competitive struggle.

The most important rights and privileges are these: professionals are generally pretty well-rewarded in income and respect. Their work is viewed as essential to the social system and so they acquire considerable influence over the way the latter is run. Most of all, they enjoy a great deal of autonomy in organizing and performing their work.

In a changing society, becoming a professional is a path of upward mobility for many individuals, and becoming a profession is a path of upward mobility for many occupational groups. By developing the basic characteristics or at least the trappings of traditional professions and by propagating a view of themselves as possessing those characteristics, the group strives to achieve professional status and its privileges.

Codes of ethics play two roles in this project. In the first place, the possession of a code is taken to be one of the essential characteristics of a true profession. The reason for this is fairly complex. Only professionals are thought to be capable of judging professional work, and even the professional's peers are reluctant to criticize his or her particular decisions in view of the complexity of the work, the judgment involved in applying general principles to concrete cases, and the confidentiality that often obtains between professional and client or employer and screens the work from external scrutiny. Professionals, as a consequence, are largely self-governing. The interests of public can be protected only if they are motivated to meet the ethical demands of society and their profession. The profession professes to determine corporately, for society as well as itself, the appropriate principles and to promulgate them through the codes of its professional associations. The application of principles to cases is left for the most part to the individual. Motivation to do so is supposed to be provided by his or her professional conscience, acquired during socialization into the profession and subsequently nurtured by its subculture. At least this is the image of the profession that publication of its codes conveys. A code of professional ethics proclaims that an occupation has moral standards and its practitioners live by them. The code, whatever its content, certifies that the occupation is a true profession—at least that is what the occupation group intends.

The actual content of the code contributes to the image of what it is to be a professional among those who assume that most professionals in fact do what the code requires. The more idealized the image, the better it serves ideological purposes—as long as it is accepted—but unfortunately it may thereby less express the operative norms of the group and move too far away from the realities of practice to shape them. Ideological self-glorification militates against effective regulation of practice.

A code that depicts the professional in ideal terms fortifies the group's claim to professional status. There are two sides to the process. Obviously the group wants to persuade outsiders whose support is needed, social and political leaders, clients and customers, and the taxpaying public. But it also must persuade itself. The loyalty and dedication of the group must be marshalled for the professional project. Professionals must take pride in their profession and think of themselves as deserving high status. The code is one instrument in the socialization process in which these attitudes are cultivated. Sometimes students in professional schools or candidates for licensing are asked to pledge fidelity to the code. Sometimes special pledges are devised for ceremonial occasions, for example, ECPD's "Faith of the Engineer," and NSPE's "Engineers' Creed," which the latter reports is "used in a number of ceremonies by the engineering profession," such as "registration certificate presentations" and "officer installation ceremonies at all levels of the Society."[37] Even if codes and creeds had no effect on the ethical level of engineering practice, they might be justified if they help generate a sense of unity among engineers and this is used to raise technical norms of competence.

This discussion of ideology reveals more clearly the ideological advantages of vagueness, which we noted early in our discussion. Vague and glittering principles in a code may persuade the public that the profession is committed to heroic measures to serve the public ethically (the myth of professional as hero). They enable professionals themselves to enjoy a glow of self-righteousness and pride in group identity. Yet no one is forced to carry a burden of onerous specific duties.

I do not mean to say that professionals are more devious or wicked than the rest of us. I only mean to call attention to the ideological utility of vagueness, given the quotient of venality in us all. Else why insist on having a code at all, if it must be too vague to regulate conduct in order to be accepted by a profession?

We must also recognize that deception of the public requires a certain amount of self-deception among people who are generally honest and sincere. A code may become an instrument of rationalization. Most professionals bring to their work what we have called ordinary conscientiousness (Brandt) and limited goodwill (Baier). They want to do what is right and are distressed if they think that doing wrong is part of their job. A set of vague principles that can be construed narrowly so as not to require much sacrifice is a shield against guilt. So can a code—which the profession assures the individual—covers all major ethical dilemmas, but which in fact has no application to his or her work environment. The individual is reinforced in the tendency to think of him or herself as a cog in a machine rather than as a moral agent.

We have touched on three uses of ethical codes as ideological instruments, as public relations documents, for ceremonies within the profession itself, and as compendia of rationalizations for the individual. These uses have obvious utility for advancing the action system that is the profession as its leaders and many of its members and patrons want it to be constituted. But social functions are not necessarily rational functions. It is obviously much harder to justify the ideological functions of codes than their regulative functions. What is more, as we have seen, ideologial functions may interfere with regulative functions. A profession then must decide what it values most, a real moral commitment to human welfare or its own special privileges. It is my impression that engineers have not yet come to grips with this issue.

Macro-Ethical Roles of Codes

The work of evaluating codes of professional ethics, and the ECPD Code in particular, has been barely begun by this paper. We have not addressed macro-ethical issues, whose resolution might cause a complete gestalt-shift in the way we view codes. John Ladd draws the relevant distinction as follows:

> Micro-ethical issues concern the personal relationships between individuals, e.g., between individual professionals and their clients, colleagues, or employers. . . . The macro-ethical problems involve questions about the social responsibilities of professionals as a group. The issues of macro-ethics arise from the tremendous *power* of professions in our society rather than professionalism as such; and power begets responsibility.[38]

We have discussed the role of ethical codes in enabling and encouraging the professional to act ethically as far as possible within the framework of the present action system that constitutes our society. We have not considered the possibility that the society itself has grave flaws such that, by serving it conscientiously, the individual incurs moral guilt or is prevented from living the most elevated moral existence imaginable.

I shall mention only two macro-ethical issues that raise questions about the morality of the practice of engineering as idealized by its leaders. I will discuss briefly the role of codes in any effort to improve the macro-ethical stance of the profession.

The first issue is whether the paternalism that is indigenous to professionalism is morally justified. As has been widely noticed by sociologists,[39] professionals typically enjoy a certain mystique by virtue of the esoteric character of their knowledge. They are viewed as having the power to solve vital problems for their clients by methods that outsiders cannot understand and are unqualified to criticize. At the same time, the profession professes a dedication to human welfare as its collectivity orientation. We noted at the outset of our discussion that the result is that professions have assumed responsibility for determining for society its norms and ends, as well as the most effective means for those ends, in their domains of competence. Individual professionals tend to do the same for their clients and publics. Admittedly this can be designated ''paternalism'' only in an attenuated sense of the word, but there is a macro-ethical issue as to whether it should be promoted as a systematic attitude. Discussions of paternalism often begin with Mill's declaration:

> . . . that the sole end for which mankind are warranted, individually or collectively, in interfering with the liberty of action of any of their number is self-protection. That the only purpose for which power can be rightfully exercised over any member of a civilized

> community, against his will, is to prevent harm to others. His own good, either physical or moral, is not a sufficient warrant.

Mill qualifies his doctrine by saying that it

> . . . is meant to apply only to human beings in the maturity of their faculties. We are not speaking of children, or of young persons below the age which the law may fix as that of manhood or womanhood. Those who are still in a state to require being taken care of by others, must be protected against their own actions as well as against external injury. For the same reason, we may leave out of consideration those backward states of society in which the race itself must be considered as in its nonage.[40]

More recently, paternalism has been defined as ''interference with a person's liberty of action justified by reasons referring exclusively to the welfare, good, happiness, needs, interests or values of the person being coerced.''[41] They key questions that must be answered to draw a line between morally unjustifed paternalism and morally justified paternalism or other morally justified treatment of others is the condition of the person whose interest is being considered—whether he or she falls into the class of children and others in a dependent state—and the sort of pressures that constitute ''coercion'' in the relevant sense. The paternalism of professionalism, if it may be so called, is not coercive in any strong sense. Clients or segments of the public usually contract freely to allow the professional to shape their ends and determine what is in their best interest, or at least they unconsciously acquiesce in this by accepting technical definitions of problems and technical formulations of the solutions that are ''reasonable.'' Nevertheless, clients come to professionals when they need help, there is the aforementioned mystique hovering about the professional, and the result may be a dependency relation that makes it difficult for the client to retain the power of decision without active encouragement from the professional.

The alternative to paternalism would be a principle of informed autonomy for the client. The ethical norm for the professional might be (a) to provide full information to the client about the latter's viable options, their chances of success, costs, and remote consequences, and (b) resolutely refuse to use his or her authority to persuade the client to do what the *professional* thinks is best.

Engineers are less likely than, say, physicians, to succumb to the temptations of paternalism on the individual level, and they are not tightly enough organized to determine corporately nuclear energy or automobile-safety policy, say, in the way physicians have determined the nature of the health-care system. Nevertheless, they doubtlessly often resort to the rationalization of paternalism, that grounds for technical recommendations simply cannot be explained to uneducated, irrational, or uninterested clients or publics.[42] Therefore, it seems worthwhile to remind engineers that there *is* a temptation of experts to use the authority of expertise to take over decisions from clients or employers and that there *is* motivation to do so in the material advantage which the enhancement of mystique brings. There is a need, in Mill's words, to raise "a strong barrier of moral conviction against the mischief" of paternalism.[43]

The ECPD Code does not do this, though it has weak implications in the right direction in its reference to "notification of proper authority" of dangers to life, safety, health, or welfare (Guidelines 1.c, 1.d, 4.h), to performing services only in the Engineer's area of competence (G.2), and to extending public knowledge of engineering (G.7.f).

The macro-ethical issue of the orientation of the whole system of professional practice toward paternalism or informed client autonomy is a complex one that needs to be explored before a final judgment can be made about the value of particular ethical codes slanted in the one direction or the other, or silent on the whole issue.

The second macro-ethical issue that I shall mention is whether a code should recognize social obligations of distributive justice for the profession. I find Frankena's formulation of the equalitarian principle of distributive justice appealing:

> Treating people equally does not mean treating them identically; justice is not so monotonous as all that. It means making the same relative contribution to the goodness of their lives (this is equal help or helping according to need) or asking the same relative sacrifice (this is asking in accordance with ability).[44]

There are difficulties in reconciling a social ethic primarily oriented toward justice with utilitarianism, which I have chosen as the point of view for this paper. I would effect the reconciliation through the principle that consciousness of belonging to a just social system and contributing to its just acts is an essential ingredient of happiness, as well as the fact that in most instances just distribution of social benefits maximizes the total benefits that are collectively enjoyed. Hence, measures, and specifically ethical rules, that promote justice have high utility.

It is clear that professionalism as a rule-governed action system is defective in this respect. By and large, professional work is differentially available according to the patron's ability to pay. There is no question that affluent classes enjoy better medical, legal, informational, accounting, and, yes, engineering services than the poor in our society. These are certainly *not* distributed according to need. The only mitigation of the condition is accomplished by public sector and eleemosynary professionalism and the limited voluntary *pro bono* contributions of individual professionals in private employment.

The professions did not invent the social system that is based on the ability to pay. They have only adjusted to it and striven for status within it, taking on many of its characteristics in the process (as have professions in other social systems in the industrial age). Thus, as we have noted, the ECPD Code assures employers that engineers are particularly loyal and competent employees and hence worthy of high responsibility and handsome compensation. The cash value of the provisions relating to "enhancing public welfare," "holding it paramount," and "improving the environment to enhance the quality of life," turns out, at most, to be an application of the principle of nonmaleficence: at least do not harm the rest of society while serving your patron. Thus, as we have seen, engineers are enjoined to inform the proper authority when they "observe conditions which they believe will endanger public safety or health" (G.1.c.3). It is true that they are also exhorted to "work in civic affairs" (G.1.e.), but it is not said to what end. No mention is made of reforming society so that the benefits of engineering services are distributed more fairly—or even of opening the ranks of the profession to disadvantaged minorities or women. The closest to this last in the official documents of the profession is a single principle in the Guidelines to Employment: 'Factors of age, race, religion, political affiliation, or sex should not enter into the employee/employer relationship." (Objective 5) (One also hopes that the fact that the ECPD Code eschews the masculine pronoun of the NSPE Code and always refers to "Engineers" in the plural reflects an effort to avoid male chauvinism.)

Obviously the addition of new provisions to ethical codes that implement the principles of client autonomy and distributive justice would not alter the ethical stance of a profession by themselves, much less correct the moral imperfections of society. These ambitious objectives would require extensive debate and ultimate consensus within the profession as to what its stance should be and a serious entrance by the profession into moral dialog with the rest of society. Noble declarations would be a minor part of the process, but they could provide a useful focus for debate.

Glossary

Acronyms of Engineering Associations

(with 1974 membership)

AACE	American Association of Cost Engineers (2300)
AAEE	American Academy of Environmental Engineers (2500)
ACEC	American Consulting Engineers Council
ACME	Association of Consulting Management Engineers (50)
AIAA	American Institute of Aeronautics and Astronautics*
AICE	American Institute of Consulting Engineers (500)
AIChE	American Institute of Chemical Engineers (38,000)*
AIIE	American Institute of Industrial Engineers (20,000)*
AIME	American Institute of Mining, Metallurgical and Petroleum Engineers*
AIPE	American Institute of Plant Engineers (6000)
ANS	American Nuclear Society (9000)*
APCA	Air Pollution Control Association (6500)
ASAE	American Society of Agricultural Engineers (7000)*
ASCE	American Society of Civil Engineers (67,000)*
ASCET	American Society of Certified Engineering Technicians (4000)
ASEE	American Society for Engineering Education*
ASHRAE	American Society of Heating, Refrigerating, and Air Conditioning Engineerings (28,000)*
ASME	American Society of Mechanical Engineers (60,000)*
ASQC	American Society for Quality Control (20,000)
ASSE	American Society of Safety Engineers (11,000)
ECPD	Engineers Council for Professional Development
IEEE	Institute of Electrical and Electronics Engineers (165,000)*
IES	Institute of Environmental Sciences (2000)
NCEE	National Council of Engineering Examiners*
NICE	National Institute of Ceramics Engineers (1700)*
NSPE	National Society of Professional Engineers (68,000)*
SAE	Society of Automotive Engineers (26,000)*
SAME	Society of American Military Engineers (23,000)
SFPE	Society of Fire Protection Engineers (1500)
SPE	Society of Plastics Engineers (17,000)
SPHE	Society of Packaging and Handling Engineers (1700)

*Members of or affiliated with ESPD.

Notes and References

[1]See William H. Wisely, "The Influence of Engineering Societies on Professionalism and Ethics," *Ethics, Professionalism and Maintaining Competence* (American Society of Civil Engineers), reprinted in Robert Baum and Albert Flores, eds., *Ethical Problems in Engineering* (Troy, NY: Rensselaer Polytechnic Institute, 1978), p. 16.

[2]Wilbert Moore, *The Professions: Roles and Rules* (New York: Russell Sage Foundation, 1970), p.3.

[3]Everett Hughes, "Professions," *The Sociological Eye: Selected Papers* (Chicago: Aldine-Atherton, 1971), pp. 41–42.

[4]Hughes, "The Study of Occupations," *Sociological Eye,* p. 398.

[5]The most able attempts to determine the phases and limits of professionalization appear to be Harold Wilensky's "The Professionalization of Everyone?" *American Journal of Sociology,* LXX (1964), pp. 137–158, and William J. Goode's "The Theoretical Limits of Professionalization," in Amitai Etzioni, ed., *The Semi-Professions and Their Organization* (New York: The Free Press, 1969). The term "professional project" is that of Magali Sarfatti Larson, who discusses its elements in *The Rise of Professionalism* (Berkeley: University of California Press, 1977).

[6]Ernest Greenwood, "Characteristics of a Profession," *Social Work,* II (1957), p. 45.

[7]"Professions and Social Structure," *Essays in Sociological Theory,* revised edition (Glencoe, IL: The Free Press, 1954).

[8]Convenient summaries of the standard model can be found in Bernard Barber, "Some Problems in the Sociology of the Professions," in Kenneth Lynn, ed., *The Professions in America* (Boston: Houghton Mifflin, 1965); Morris L. Cogan, "Toward a Definition of Profession," *Harvard Educational Review,* XXIII (1953), pp. 33–50; Greenwood, "Characteristics of a Profession"; Hughes, "Professions"; Geoffrey Millerson, *The Qualifying Associations* (London: Routledge and Kegan Paul, 1964); and Ronald M. Pavalko, *Occupations and Professions* (Itasca, IL: F. E. Peacock, 1971).

[9]Wisely, *op. cit.*

[10]Wisely, *op. cit.,* pp. 15–16.

[11]Jane Clapp, ed., *Professional Ethics and Insignia* (Metuchen, NJ: The Scarecrow Press, 1974), p. 246.

[12]See note 8 for references to discussions of the model.

[13]Wilbert Moore, *op. cit.,* p. 51.

[14]I have addressed this issue in "Professional Ideals and Ideology," in Robert Baum and Albert Flores, eds., *Ethical Problems in Engineering,* 2nd ed. (Troy, NY: Center for the Study of the Human Dimensions of Science and Technology, 1980).

[15]These figures were taken from Clapp, *Professional Ethics and Insignia.* They hold for 1974 or before and are only approximate. I have also relied on Clapp for other data about engineering associations. It should be noted that she appears to take self-descriptions provided by the associations at face value.

[16]Clapp, *op. cit.,* p. 260.

[17]William J. Rothstein, "Engineers and the Functionalist Model of the Professions," in Robert Perucci and Joel E. Gerstl, eds., *The Engineers and the Social System* (New York: Wiley, 1969), reprinted in Robert Baum and Albert Flores, eds., *Ethical Problems in Engineering* (Troy, NY: Center for the Study of the Human Dimensions of Science and Technology, 1978), p. 18.

[18]See Wisely, "The Influence of Engineering Societies," and Rothstein, "Engineers and the Functionalist Model of Professions."

[19]For the most part, these are the actual and imaginary petitioners seeking ethical advice in Philip L. Alger, N. A. Christensen, and Sterling P. Olmsted, *Ethical Problems in Engineering* (New York: Wiley, 1965) and the NSPE's publication, *The Opinions of the Board of Ethical Review.*

[20]See Richard B. Brandt, "Toward a Credible Form of Utilitarianism" in Hector-Neri Castenada and George Nakhnikian, eds., *Morality and the Language of Conduct* (Detroit, MI: Wayne State University Press, 1963), and "Some Merits of One Form of Rule-Utilitarianism," *University of Colorado Studies, Series in Philosophy,* No. 3 (Boulder: University of Colorado Press, 1967). The formulas used here are adapted from the latter article.

[21]Brandt's more exact (and more complex) definitions are the following: (1) An act is right if and only if it would not be prohibited by the moral code ideal for the society; and an agent is morally blameworthy (praiseworthy) for an act, if, and to the degree that, the moral code ideal in that society would condemn (praise) him for it. (2) A moral code is ideal if its currency in a particular society would produce at least as much good per person (the total divided by the number of persons) as the currency of any other moral code. (3) For a moral code to have currency in a society, a high proportion of the adults in the society must subscribe to the moral principles, or have moral opinions, constitutive of the code, and principles belong to the code only if they are recognized as such by a large proportion of adults. "Some Merits of One Form of Rule-Utilitarianism," p. 48.

[22]I am not entirely happy with the terms "social function," "rational function," "sociological understanding," and "philosophical evaluation," but there is no standard terminology for what I have in mind and these terms do suggest some of the connections as well as distinctions among the concepts for which they stand.

[23]Brandt, *op. cit.*

[24]Daniel W. Mead, *Manual of Engineering Practice,* No. 21 (American Society of Civil Engineers), in Baum and Flores, *op. cit.,* p. 26.

[25]In an unpublished paper.

[26]Stephen H. Unger, "Engineering Societies and the Responsible Engineer," *Annals of the New York Academy of Sciences,* CXCVI, 10 (1973), p. 434.

[27]The feasibility and desirability of guaranteed freedoms for employees in general has been explored at length by David Ewing in *Freedom Inside the Organization* (New York: McGraw-Hill, 1977). One author's idea of what these might include for engineers is to be found in Robert L. Whitelaw, "The Professional Status of the American Engineer: A Bill of Rights," *Professional Engineer* (August, 1975), pp. 37–41, reprinted in Baum and Flores, *op. cit.,* pp. 68–70.

[28]NSF Summer Workshops on Ethical Issues in Engineering, Illinois Institute of Technology, 1979.

[29]Reprinted in Baum and Flores, *op. cit.,* pp. 77–81.

[30]"Toward a Credible Form of Utilitarianism."

[31]Kurt Baier, "Responsibility and Action," in Myles Brand, ed., *The Nature of Human Action* (New York: Scott Foresman, 1970).

[32]Baier, "Responsibility and Action."

[33]John Stuart Mill, *Utilitarianism* (New York: Library of Liberal Arts, 1957), p.25.

[34]In "Abstract of Remarks to Workshop on Professional Ethics: The Role of Scientific and Engineering Societies, of the American Academy for the Advancement of Science."

[35]"The Theoretical Limits of Professionalism," pp. 277, 279.

[36]It is possible to adapt the idea of political ideology to this context and speak of professional ideologies since the effort of occupational groups to professionalize them-

selves has a political dimension, in the broad sense of "politics" as the effort to shape the structure of social institutions. We are dealing here with what John Plamenatz calls a "partial ideology" in *Ideology* (New York: Praeger, 1970), p. 18. A partial ideology is used by a group to gain a position, partly by political pressure, in the existing society, in contrast to "total ideologies," which are platforms on the basis of which groups seek to gain or retain control of the state and introduce, or preserve, an entire social order. The notion of professional ideology is put to good use by Edwin Layton, in *The Revolt of the Engineers* (New York: Harper and Row, 1973), and Margali Sarfatti Larson, in *The Rise of Professionalism.*

[37]Clapp, *Professional Ethics,* p. 251.

[38]Abstract.

[39]See Moore, *The Professions,* pp. 32–33 on the professional as shaman; and Hughes, *The Sociological Eye,* p. 318, on the "charisma of skill."

[40]John Stuart Mill, *On Liberty,* Chapter I.

[41]Gerard Dworkin, "Paternalism," in Richard Wasserstrom, ed., *Morality and the Law* (Belmont, CA: Wadsworth, 1971), pp. 144. See this article and Joel Feinberg, "Legal Paternalism," *Canadian Journal of Philosophy* I (1971), for explorations of issues connected with legal paternalism. They draw a number of distinctions that are useful for evaluating professional paternalism, particularly as it is reflected in state licensing.

[42]Thus Norman Bowie at a recent conference asserted that the basis for paternalism among professionals is epistemological, not moral.

[43]*On Liberty,* Chapter I.

[44]William K. Frankena, *Ethics* (Englewood Cliffs, NJ: Prentice-Hall, 1963), p. 41.

Professional Autonomy and Employers' Authority

Mike W. Martin

Need for establishing authority arises in circumstances where unrestrained individual discretion in decision-making conflicts with a desired degree of social organization and order. Accordingly, submitting to authority typically involves a willingness to accept some policies and directives that on their own merits one may find inadequate. Even to submit to the authority of a basketball referee requires playing by occasionally bad or even outrageous rulings. Similarly, an employee who regards the employer as having legitimate authority over him or her acknowledges the necessity to adhere to at least some of the employer's orders and the company's regulations whose rationale may be found wanting. Doing so voluntarily is part of being a faithful agent and trustee of the employer. Presumably this holds true for the many areas of business operations that affect the safety, health, and welfare of the general public. Many employees, however, are also members of a profession. As such, they are generally regarded as obligated to exercise their independent skilled judgment in a manner they calculate will avoid harm and promote the good of the public.

It might seem then that the roles of the professional and of the faithful employee are strictly incompatible, and not simply in the commonly understood sense that a person functioning in both roles will occasionally face competing obligations. The point is rather that in matters pertaining to the public good it seems *logically* impossible to function simultaneously in both roles. Professionalism demands exercising autonomy in forming and acting on an independent view of what causes harm to the public or promotes the positive good of the public. Fidelity to an employer demands foregoing autonomy in submitting to the directives of an employer—directives that are based on the employer's view of what the public good is and what the organization owes the public.

There is a conceptual problem here—what I will call the Reason-Compatibility Problem—that arises because the types of grounds for acting in matters concerning the public good required of the morally sensitive professional and the loyal employee seem incommensurable. Technically, the same problem would arise even with respect to an imaginary world in which by good fortune employers and professionals always agree in their judgments. For even in that world there would remain a difference in reasons for action: apparently the loyal employee would act because the boss directed an act, while the professional would act because his or her independent judgment led to the conclusion that the same act was best for the public. Resolving this conceptual problem will not remove the practical moral dilemmas faced by employed professionals, but it does promise an enriched understanding of the nature of those dilemmas and possibly some general guidelines for action.

I will focus on the profession of engineering, for there the problem is unusually poignant. Owing to the needs of modern technological bureaucracies, roughly ninety percent of engineers are salaried employees. Yet engineering codes of ethics explicitly prescribe as obligatory a degree of independence that seems incompatible with an engineer operating as a loyal employee—a requirement that the codes also prescribe as obligatory. Within this context, I will attempt partially to resolve the Reason-Compatibility Problem by arguing against the following two extreme positions, each of which gives rise to the problem: (1) Loyalty to employers and recognition of their authority involves accepting directives without any critical review of them in light of professional and moral considerations; (2) The professional's moral obligations to the public require that all professional activities conform to his or her own direct calculation of their consequences for the public good.

The first view is sympathetically set forth by Herbert Simon in his classic text, *Administrative Behavior*. A subordinate accepts the superior's authority, according to Simon, when he or she "holds in abeyance his own critical faculties for choosing between alternatives and uses the formal criterion of the receipt of a command or signal as the basis for choice."[1] Not only does this mean that the subordinate abdicates choice based on personal reasoning regarding the best course of action, but that the superior's suggestions and orders are accepted "without any critical review or consideration."[2] At most, the subordinate's reasoning is aimed at anticipating commands by asking how the superior would wish him or her to behave in the given circumstances.[3] Simon emphasizes that all employees—even the most submissive ones—place limits on the "zone of acceptance" in which they submit to their employer's authority. But within that zone Simon portrays the loyal employee as "relaxing his own critical faculties," and permitting "his behavior to be guided by the decision of a superior, without independently examining the merits of that decision."[4] The justification for such recognition of authority lies in the benefits deriving from its key role in securing coordinated group behavior. Specifically, as backed by sanctions, it maximizes the responsibility of the employee to the employer and helps assure a high degree of efficiency and of productivity.[5]

It is clear that an employed engineer cannot recognize the employer as having authority in the way Simon describes. A professional is bound by ethical norms that must guide all dealings with an employer. One can recognize no zone of acceptance in which one suspends all critical scrutiny of the employer's directives in light of professional obligations. There must always be a moral ground outside the confines of business dealings on which one is willing to take a stand in assessing all the employer's directives. Blind obedience is too high a price to pay for increased social organization. But having said this, the question remains about the nature of the norms in light of which one must guide all business dealings. Are they grounded always and solely in a personal assessment of the public good? This leads us to the second extreme view, which holds that all of the professional's activities must conform to his or her own direct calculation of the public good.

This view is prescribed for engineers by many of the codes of ethics of their professional societies, at least on one literal interpretation of them. Consider, for example, the code set forth in 1974 by the Engineers Council for Professional Development, an umbrella organization for numerous engineering societies, most of which endorse the code.[6] The code opens with the pronouncement that "Engineers shall hold paramount the safety, health and welfare of the public in the performance of their professional duties." It then explains that this is to be understood in a strong sense: "Engineers shall not approve nor seal plans and/or specifications that are not of a design safe to the public health and welfare." Moreover, if their professional judgment is overruled "where the safety, health, and welfare of the public are endangered," they must "inform their clients or employers of the possible consequences and notify other proper authority of the situation as may be appropriate." The final clause constitutes an injunction to whistleblow outside the organization whenever the engineer judges that the public interest is being harmed by the employer and he or she has no further recourse within the organization. One writer has claimed in this same spirit that until engineers have the guaranteed rights to act in the way prescribed by the codes, free of all employer coercion, there is no justification for calling them professionals.[7]

In a recent response to the ECPD code, Samuel Florman has argued that if taken seriously by most engineers it would lead to nothing short of chaos:

> Ties of loyalty and discipline would dissolve, and organizations would shatter. Blowing the whistle on one's superiors would become the norm, instead of a last and desperate resort. It is unthinkable that each engineer determine to his own satisfaction what criteria of safety, for example, should be observed in each problem he encounters.[8]

Even after allowing for the apocalyptic hyperbole, Florman's words are sobering. Encouraging each engineer to act solely on his or her own personal assessment of the public good would be sufficient to destroy the degree of coordination essential to the effective functioning of modern bureaucracies. To be sure, Florman lapses too far in the direction of Simon's model when in places he treats engineers' moral duties to the public as fulfilled in their role as private citizens, suggesting that engineers should not "filter their everyday work through a sieve of ethical sensitivity."[9] Surely such filtering is precisely what is required of professionals, and the only question concerns the nature of the sieve. But Florman is correct to emphasize that an engineer must submit at least partially to the direction of the employer in matters concerning the public good.

Both of the two extreme views, then, must be rejected. On the one hand, Simon's model of what it is to accept authority is incompatible with being a professional or indeed with being a moral agent. On the other hand, acting solely on one's independent judgment of what best promotes the public welfare is incompatible with being a faithful employee within an efficiently operating formal organization. The correct view lies somewhere in between the extremes. Submitting to an employer's authority does not mean abandoning one's general responsibilities as a moral agent and one's special responsibilities as a professional to assess the moral merits of an employer's orders. At the very least, engineers and other professionals are responsible for forming and expressing within their organization their own independent views on the impact of projects on the public good. And where the danger becomes extreme, the engineer must be prepared to take further recourse outside the organization. Nevertheless, submitting to an employer's authority does involve giving special weight to the employer's directives, a weight that must be set against the professional's autonomous calculation of the independent merit of the act directed by the employer. Assuming that the organization meets certain minimal standards of justice, the employee-employer relationship can create a distinct moral obligation that prevents the professional from acting always and solely on an independent personal assessment of the public good. Though one must continuously exercise independent *judgment* about the consequences of one's work for the public good, one cannot be expected in one's *actions* to disregard altogether the organization's directives concerning the "safety, health, and welfare" of the public.

It may be objected that all I have said is that employee status prevents one from being a full-blown professional and fulfilling one's moral obligations as a professional. That is, that the Reason-Compatibility Problem remains unresolved, for the professional's obligations entail making all of his or her work activities conform to an independent calculation of the consequences of those activities for the public good. It is difficult to respond to this objection without providing a fuller analysis of the concepts of professionalism and professional autonomy than the scope of this essay allows—that is one reason why I earlier spoke of providing a 'partial' resolution of the problem. As a brief reply, we must first insist that any such analysis take account of most of the salient features of the paradigm professions. One of these features is that a purely act-consequentialist form of reasoning is ruled out by many central professional norms. That is, in certain major respects it is held to be wrong for a professional to base his or her actions solely on a direct personal calculation of the consequences of those actions. A lawyer, for example, is forbidden from freely expressing information about a client's private testimony even though the lawyer may frequently believe it will have the best consequences in the situation. A doctor cannot force an operation on a patient simply because he or she thinks it is best for the patient and others involved. The general practices of client confidentiality and patient autonomy are widely believed to be justified by their producing generally good consequences (as well as by according patients their rights). But it is not assumed that each act must be justifiable in terms of an appeal to consequences.[10]

Similarly, loyalty to an employer can be viewed as part of the norms governing the professional activities of many people. As a general principle, it can be justified in terms of the public benefits deriving from coordination and efficiency in production that such ties contribute to. It can also be argued that ties of loyalty are capable of humanizing the employee-employer relationship in a way that reduces the alienation of purely money-based relationships. But each act of obedience need not itself be justified solely in terms of good consequences for the public. Within limits, the employee obligation may on occasion carry an importance that outweighs the results of a direct calculation of the public good. (Although not always—even client confidentiality must occasionally be violated when extremely harmful consequences are involved.[11])

These conclusions will perhaps readily be granted with respect to the public "welfare." But the term is notoriously ambiguous, as well as vague, and has at least the following very different senses: what the majority of people affected desire or want; what they would want if they were well-informed; what is objectively good for them, whether they want it or not; what best satisfies the rights and legitimate claims of those affected. Moreover, the influence of engineering products on the public's welfare and interest includes such things as a product's esthetic features, the average durability of the product and its parts, energy-efficiency, and the impact on the beauty of the environment. Though the imput of engineers on all these matters is essential, most will concede that it is generally proper for engineers to be willing to work according to the organization's final view. But how about with respect to the public health and safety?

First, is it ever permissible for an engineer—*qua* professional with obligations to the public good—to submit to an employer's orders to undertake a project he or she believes too risky in terms of public health and safety? We are all aware of how a personal moral obligation to one's family may on occasion require undertaking work that one would otherwise disapprove of, but is there ever a professional basis for such work? The two extreme views would have us answer, respectively, "Of course—anytime!" and "Of course not—never!." However uninspiring, the correct answer is "Sometimes yes and sometimes no." Suppose an engineer is assigned to work on a new airport facility that has been approved on the basis of the same hard facts known by the engineer and by the vast majority of voters affected by the project. The engineer may disagree with the voters and feel that the health hazards that will arise from noise and air pollution are too great to warrant the project, but he or she need not be acting immorally in acceding to the employer's orders to work on the project.

Robert Baum has convincingly argued that such cases are closely analogous to those where a patient has the right to make the final decision about whether to undergo an operation. After all, a judgment of safety is a judgment of acceptable risk, and within limits the user of a product has the right to make a personal value judgment concerning the acceptability of risk probabilities (assuming he or she is capable of doing so).[12] Baum, however, goes too far when he avers, "It is especially important to recognize that doing something that one believes is harmful is not morally wrong, if the parties most affected want it done."[13] This view would open the door to justifying atrocities on masochistic consenters and wholesale disruption of the professional's inner integrity. It is often morally permissible for a doctor to perform an operation chosen by an informed patient even though the doctor views the operation as far too risky and hence harmful to the patient and the family. But there are limits: it is patently wrong for a doctor to work in an abortion clinic if he or she views most abortions performed there as the murder of innocent lives, even though the abortions are both legal and chosen by an informed patient. Similarly, it is immoral for an engineer to participate in building nuclear weapons if he or she is convinced that the sum of such activities is soon going to bring an end to humanity.

Second, is it ever morally permissible for an engineer, *qua* professional, to conduct a project in a way that he or she believes goes against an adequate degree of public health and safety? The tempting extreme views, with their comfortable aura of simplicity and certainty, would have us answer with an unqualified "yes" or "no." But again there are cases and cases. The professional must always seek to work at least according to standards of accepted engineering practice. But what about instances where an engineer's autonomous judgment suggests that those standards are not high enough to insure what he or she sees as an adequate degree of public safety? Suppose, for example, that an automotive engineer believes that the accepted safety standards for car bumpers are grossly inadequate. I believe that after expressing his or her view, the engineer is justified in following the employer's design specifications, while working independently to upgrade the standards accepted by the profession, the employer, and indeed the public.

The cases I have cited are special in that the expressed desires of the client, the majority of the public, and the group using and consenting to the use of the product are roughly the same. It is not possible here to discuss other cases where the condition is not met, but one general comment is in order. The moral violations that justly outrage us are those in which an employer orders an engineer to collaborate in withholding essential safety information from a client, the public, or special groups affected. An employer orders silence concerning known defects in the design of a passenger jet's cargo door,[14] the design of wheels of a fighter jet,[15] or the construction of a nuclear power plant.[16] No justification in terms of fidelity to the employer can be given for blatant disregard of human life. In such cases every legal, psychological, and financial support must be given by professional societies, colleagues, and an aware public to the conscience-guided public guardian who whistleblows.

I have argued that the Reason-Compatibility Problem of the apparent logical impossibility of being both a loyal employee and a genuine professional in matters concerning the public good is partially resolvable. The problem arises in a general form only when the employed professional is mistakenly viewed as obligated to act with respect to the public good solely on the basis of either his employer's orders or his own independent judgment. The moral status of a professional who is a loyal employee in many ways parallels that of a moral agent who is a loyal citizen. Both are sometimes confronted with orders or regulations they do not view as promoting the public's good. Both are obligated: (a) to identify such regulations by regularly forming an independently reasoned view concerning the public good, (b) to express that view freely either within the general limits of responsible speech or the organization's limits, (c) to seek to promote free expression of responsibly formed views by creating an atmosphere of tolerance and liberal channels of communication within organizations, (d) to be willing at some point to engage in civil disobedience or organizational disobedience (e.g., whistle-blowing) when confronted with seriously immoral laws or orders creating extreme danger to the public for which there is no other effective remedy.[17] Both must act from two types of reasons: respect for legitimate authority and concern for the public good. Attempting to reduce either's proper reason-base to just one of these leads to unacceptable extremes of sanctioning blind obedience or denying the justification of any authority in matters pertaining to the public good.[18]

Properly conceived, the employed professional's duty to exercise autonomy lies not in basing his or her actions solely on one type of reason, but in constantly finding a responsible way to reconcile or adjudicate between reasons of both types. He or she must not only continuously form independent judgments about what is best for the public, but weigh those judgments against an employer's goals. No practical guidelines, of course, any more than any simplistic theoretical position, will ever remove the possibility for these two sorts of reasons to clash in a way that forces the professional to make a painful decision one way or the other. Alongside and interwoven with the "existential pleasures of engineering," so delightfully portrayed by Samuel Florman in his book of that title, resides the anguish of being condemned to both freedom and responsibility, an anguish portrayed by earlier Existentialist philosophers.

Acknowledgments

Financial support for writing this paper was provided by NEH in funding my participation in the National Project on Philosophy and Engineering Ethics. I am grateful to my colleague on the project, Roland Schinzinger, for many helpful discussions.

Notes and References

[1]Herbert A. Simon, *Administrative Behavior,* 3rd Edition (New York: Free Press, 1976), pp. 126–127.

[2]*Ibid.*, p. 228.

[3]*Ibid.*, p. 129.

[4]*Ibid.*, p.151; p. 11.

[5]*Ibid.*, pp. 134–138.

[6]Printed in *Ethical Problems in Engineering,* 2nd Edition, Vol. 1, Albert Flores, ed., (Troy: Rensselaer Polytechnic Institute Press, 1980), pp. 65–69.

[7]Robert L. Whitelaw, "The Professional Status of the American Engineer: A Bill of Rights," *Professional Engineer* (August, 1975), pp. 37–41. Reprinted in Flores, *op. cit.*, pp. 68–70.

[8]"Moral Blueprints," *Harper's* 257 (October, 1978), p. 32. Cf. Chapter 3 of Florman's earlier book *The Existential Pleasures of Engineering* (New York: St. Martin's Press, 1976).

[9]*Ibid.*

[10]For this general distinction see John Rawls, "Two Concepts of Rules," *Philosophical Review,* 64 (1955).

[11]Cf. Leo Cass and William Curran, "Rights of Privacy in Medical Practice," reprinted in *Moral Problems in Medicine,* Samuel Gorovitz *et al.*, eds. (Englewood Cliffs: Prentice-Hall, 1976), pp. 82–85.

[12]Cf. William Lowrance, *Of Acceptable Risk* (Los Altos: Kaufman, 1976), Chapter 3.

[13]Robert Baum, "The Limits of Professional Responsibility," in Flores, *op. cit.*, pp. 48–53.

[14]Paul Eddy et al., *Destination Disaster* (New York: Quadrangle, 1976).

[15]K. Vandivier, "Engineers, Ethics and Economics," *Conference on Engineering Ethics* (New York: American Society of Civil Engineers, 1975), pp. 20–24.

[16]"Carl W. Houston and Stone and Webster," *Whistleblowing,* Ralph Nader et al. (New York: Grossman Publishers, 1972), pp. 148–151.

[17]For an analysis of the concept of "organizational disobedience" see James Otten, "Organizational Disobedience," in Flores, *op. cit.*, pp. 182–186.

[18]For an attempted defense of the latter denial with respect to state authority, see Robert Paul Wolff, *In Defense of Anarchism* (New York: Harper and Row, 1976).

Freedom of Expression in the Corporate Workplace

A Philosophical Inquiry

Robert F. Ladenson

In a novel entitled *Scientists and Engineers: The Professionals Who Are Not* the author, Louis V. McIntire, presents a highly negative picture of life as an employee in a large private business corporation.[1] Characters in the novel inveigh against management favoritism, cheating inventors out of bonuses, and taking unfair advantage of employees in employment contracts. The fictional employer bears a striking resemblance to DuPont, the company for which McIntire worked as a chemical engineer from 1956 through 1971, the year the book was published. In 1972 McIntire was fired.[2]

Should employees of large private business corporations be free to speak out on any subject without fear of dismissal or other sanctions even when they level harsh criticism at their corporate employers? In this paper I will argue that they should for reasons that closely parallel one of the fundamental bases for the principle of freedom of expression pertaining to the relation between individuals and the state. An important and controversial consequence of this view is that corporate employees should be free to speak without fear of sanctions even when they make false allegations that lead to a decline in either productivity or profits.

Before proceeding, however, the above thesis must be distinguished from another with which it can be easily conflated—that new rights to free expression for corporate employees should be acknowledged as a matter of law. The argument for this claim requires that one do more than make out a case for the desirability of freedom of expression in the corporate workplace. It also must involve presenting either a new interpretation of the First Amendment, according to which it applies to corporate employees, or a plausible case for the feasibility of protecting employee rights to free expression in the private corporate sector through a statute, or new Constitutional amendment, or through extending traditional common law causes of action to facilitate lawsuits by dismissed employees. In this paper, I will not discuss any of these important matters.[3] It seems to me, however, that before turning one's attention to them a convincing account must be put forward of why freedom of expression for corporate employees is important. Without such an account the other matters are not worth pursuing.

To place the issue in its proper perspective, a few words must be said about the current situation of corporate employees in regard to free expression. For the most part, the common law doctrine of "employment at will" governs employer-employee relations in the private sector.[4] This doctrine looks upon employee and employer as equal partners to an employment contract. Just as employees may resign whenever it pleases them, so also employers may dismiss their employees whenever they desire.

This latter aspect of the doctrine has been stated forcefully time and time again in various court decisions. For example, in *Payne v. Western and A.R.R.*. [81 Tenn. 507, 519–20 (1884)] the court declared that employers "may dismiss their employees at will . . . for good cause, for no cause, or even for cause morally wrong, without thereby being guilty of legal wrong." Similarly, in *Union Labor Hospital Association v. Vance Redwood Lumber Co.* [18 Cal. 551, 555 (1910)] the court said that the "arbitrary right of the employer to employ or discharge labor is settled beyond peradventure." The traditional doctrine of employment at will was recently invoked by the Supreme Court of Pennsylvania to dispose of *Geary v. United States Steel Corporation* [456 Pa. 171, 319, A.2nd 174 (1974)]. In this case, Geary, an employee, charged that he was unjustly dismissed by United States Steel after he went outside normal organizational channels to warn a vice president of the corporation (it turned out correctly) about defects in steel tubing that was about to be marketed.

The absence of legal rights to free expression for corporate employees would not really matter if business corporations generally tended to be places in which employees feel free to express their beliefs and attitudes. The almost unanimous testimony of people with substantial work experience in the corporate world, however, is to the contrary. When CBS televised a program on Phillips Petroleum Corporation, William V. Keeler, its chief executive said of the employee who deviates from unwritten company rules about dress, manners, or other behavior, "The rest of the pack turns against him."[5] In his review of the program in the *New York Times,* John J. O'Connor noted that at Phillips there was a "direct ratio between the extent of an individual's ambitions and the pressure for conformity."[6] Phillips may be an extreme case, but if so it is an extreme case of a situation common to most large business corporations—namely, the absence of an open atmosphere conducive to free expression by employees.

But why should there be such an atmosphere in corporations? The principle of freedom of expression, as it has been traditionally defended by political philosophers, applies solely to the relation between citizens and the state.[7] That is to say, the proposition that people ought to be free to express themselves has been taken exclusively as a proscription of governmental interference with their doing so. By contrast, interferences with expression that stem from other sources have not been regarded as falling under the purview of the principle. For example, suppose one private individual interferes with attempts at expression by another. Although such interference might warrant condemnation for a variety of reasons depending upon the circumstances—e.g. rudeness, unfairness, and so on—the situation does not count as one in which the principle of freedom of expression, as traditionally conceived of, has been violated. Some writers stress the respects in which large private business corporations resemble governments.[8] The crucial question, however, is whether they resemble them in precisely those respects that matter from the standpoint of well-entrenched philosophical defenses of the principle of freedom of expression. One cannot then move directly from such defenses to the claim that employees in private business corporations should have extensive freedom to express themselves. More in the way of analysis is needed.[9]

One can begin such an analysis by contrasting two basically different ways of making out the case for freedom of expression in corporations, which I will refer to respectively as the volunteer public guardian and the fundamental liberty approaches. The first mode of argument, the volunteer public guardian approach, sees the case for free expression of corporate employees as having to do primarily with associated potential benefits to society from increased exposure of corporate corruption, waste, and negligence. The following quotation from *Where The Law Ends* by Christopher Stone exemplifies this approach.

> . . . anyone concerned with improving the exchange of information between the corporation and the outside world must pay serious regard to the so called whistleblower. The corporate work force in

> America, in the aggregate, will always know more than the best planned government inspection system we are likely to finance. Traditionally workers have kept their mouths shut about "sensitive" matters that come to their attention. There are any number of reasons for this, ranging from peer group expectations, to the employee's more solid fears of being fired . . . .
>
> This means that if ethical whistleblowing is to be encouraged some special protections and perhaps even incentives will have to be afforded the whistleblower."[10]

The second mode of argument for freedom of expression of corporate employees, the fundamental liberty approach, does not focus upon the immediate social benefits to be gained as a result of a more open atmosphere in corporations. Instead, it suggests that we should look upon freedom of expression in the corporate workplace as an inherent right grounded in basic principles of social morality. Such an outlook is reflected in the quotation below from David Ewing's book *Freedom Inside The Organization*.

> A classic formulation of the philosophy of the First Amendment was given decades ago by Supreme Court Justice Louis D. Brandeis. Although he was commenting upon free speech in the political area, his observations would seem to be equally valid for the governance of corporations . . . Brandeis wrote: "Those who won our independence knew that . . . it is hazardous to discourage thought, hope, and imagination . . . They eschewed silence coerced by law—the argument of force in its worst form . . ."
>
> . . . many executives in business and government find (the above view) . . . "unrealistic" when it comes to employee speech. In the name of discipline, they feel that free thinking about an organization's policies should be suppressed. In this respect, if no other, they are in league with radical left philosopher Herbert Marcuse who argues that free speech cannot be justified when it becomes too distracting.[11]

We have then two kinds of arguments in support of freedom of expression in corporations. The volunteer public guardian approach stresses immediate benefits to society that will flow presumably from making the climate in corporations more conducive to free speech. The fundamental liberty approach, on the other hand, looks to basic principles of social morality akin to those that underlie the First Amendment in its most familiar applications. These two kinds of arguments differ in an important way that is brought out sharply by considering the question "What should happen to whistle blowers who turn out to be wrong?"

Following the volunteer public guardian approach one would treat this question by performing a comparative analysis of the social benefits and costs associated with corporate whistle-blowing. As mentioned above, on the benefit side one can cite the increased exposure of corporate waste, corruption, and negligence. On the cost side, however, one must include the possibility of a general decline in productivity stemming from decreased efficiency as a result of disruptions in the corporate decision-making and administrative routines. In addition, where whistle blowers are mistaken in their allegations about the safety or quality of a product the affected corporations may unfairly suffer a decline in profits.

A social cost-benefit analysis of corporate dissent not only requires attaching weights to the above factors, but also necessitates an assessment of both the prevalence of antisocial corporate behavior and the nature of its consequences. A person who regards such behavior not only as commonplace, but also as gravely harmful would advocate extensive protection for corporate dissenters, holding that the costs associated with mistaken allegations they might make count for relatively little in the balance. On the other hand, if serious corporate misbehavior is looked upon as the exception rather than the rule then a different view of the matter becomes appropriate. Indeed, depending upon how exceptional one regards it, and upon how heavily one weighs the costs associated with corporate dissent, it might be reasonable to suggest that such dissent should be thought of on analogy with the common law rules in regard to citizen's arrests. Specifically, a person making a citizen's arrest avoids tort liability for unlawful detention only if the person he or she arrested *actually* committed a felony. Reasonable belief is not a defense.[12] By analogy, someone who regards corporate misconduct as exceptional might say that freedom of expression in corporations should only extend to dissenters who turn out to be right.

The prevalence of serious corporate misbehavior, and the nature of its local consequences, are empirical issues lying beyond thc scope of this paper.[13] The point to be noted here, however, is that when one makes the case for freedom of expression in corporations by way of the volunteer public guardian approach, the question of *how much* freedom corporate employees should have involves a weighing of costs and benefits that essentially depends upon one's beliefs about these empirical matters.

By contrast, the fundamental liberty approach eschews appeal to any such considerations. If freedom of expression, conceived of as a fundamental liberty, extends to the employee-employer relationship in a corporation then questions about its nature and scope cannot be settled through balancing immediate social benefits and costs. As Ronald Dworkin has pointed out, regarding a liberty as fundamental involves believing that it "trumps" all other considerations, even those that otherwise would be considered decisive.[14] Moreover, if a coherent philosophical account of the principle of freedom of expression pertaining to citizens and the state can be extended reasonably to cover the relationship between employers and employees, then we have a short answer to our question about the whistle blower who turns out to be wrong. Such an individual can no more justifiably be made subject to sanctions by his or her employer than a citizen can justifiably be punished at the hands of the government simply for expressing incorrect views.

A crucial question for the fundamental liberty approach then is whether such an extension can be made. This question, in turn, requires a brief review of some important points about freedom of expression. To begin, the primary task for a philosophical defense of it can be stated in the following way. Acts of expression can, at times, lead to very undesirable consequences, consequences that when caused in any other way would be regarded as so grave that the behavior causing them ought to be legally prevented. Nonetheless, for those who regard the right to freedom of expression as fundamental, even when its exercise leads to certain of these undesirable consequences, limitations upon freedom of expression are still considered unjustifiable. How one can defend such an outlook must be explained.

The search for such an explanation inevitably leads to the arguments advanced respectively in chapters two and three of John Stuart Mill's classic essay *On Liberty*. Boiled down to essentials, Mill contends that countenancing routine governmental interference with the expression of beliefs and attitudes by citizens would only make sense if we believed it possible to identify infallible, perfectly benevolent human beings and to put them into positions of political power. Since, of course, this cannot be done, it follows that if governmental authorities routinely prevent the expression of beliefs and attitudes on the basis of their content, the result will be inevitably a widespread acceptance of seriously erroneous viewpoints. What is worse, this benighted condition of society will persist in all likelihood over many generations because the most obvious means of overcoming it, free discussion, will not be available. The right to freedom of expression, conceived of as ruling out governmental restrictions upon the content of beliefs and attitudes that may be expressed, can thus be treated as fundamental in view of the extraordinary social interest its acknowledgment serves—namely, the avoidance of social action predicated upon mistaken beliefs over the long run.

The foregoing argument constitutes a formidable case for freedom of citizens from governmental interference to express their beliefs and attitudes. It does not, however, apply in an obvious way to the relations between corporate employees and their employers. The argument calls attention to the grave long-run social harm that stems from giving a *single* individual or group power to regulate expression and hence to control thought. Now although corporate employers can, and undoubtedly often do, exercise substantial coercive force to discourage their employees from freely expressing themselves, it would seem that no one corporation could exercise the kind of centralized power to control thought of which a strong government would be capable. Accordingly, Mill's argument in *On Liberty* does not serve to establish that freedom of expression should exist in private business corporations.

The situation is quite different, however, with regard to Mill's line of reasoning in the third chapter entitled "Of Individuality As One Of The Elements Of Well Being." To grasp the essentials of this argument one must first concentrate upon the passage below.

> . . . to conform to custom merely *as* custom does not educate or develop in (a person) any of the qualities which are the distinctive endowment of a human being. The human faculties of perception, judgment, discriminative feeling, mental activity, and even moral preference are exercised only in making a choice. He gains no practice in discerning or in desiring what is best . . . .
>
> He who lets the world, or his own portion of it choose his plan of life for him has no need of any faculty other than the ape-like one of imitation. He who chooses his plan for himself employs all his faculties. He must use observation to see, reasoning and judgment to foresee, activity to gather materials for decision, and when he has decided, firmness and self-control to hold to his deliberate decision. And these qualities he requires and exercises exactly in proportion as the part of his conduct which he determines according to his own judgment and feelings is a large one.[15]

In chapter three of *On Liberty* Mill can thus be thought of as arguing in the following way. Certain abilities and capacities, such as observation, judgment, discrimination, firmness of will, and so forth are the distinctive endowment of a human being. These abilities and capacities, which Mill takes to be the elements of what he terms "individuality," make it possible to discern and desire what is best. Thus, in the proportion to which people have them, they become both more valuable to themselves and potentially more valuable to others.[16] According to any reasonable conception of the good for society, it should be a primary function of social arrangements to facilitate everyone's cultivating his or her individuality, as understood above, to the greatest possible degree. Individuality, so understood, however, consists in the possession of a variety of different abilities and capacities, all of which can only be developed by exercising them. Without freedom of expression, however, the likelihood for such development on a large scale is extremely low. Accordingly, even if freedom of expression sometimes leads to serious harm, this must be borne as a cost of making it possible for a society to develop in which large numbers of people cultivate their individuality.[17]

Unlike the line of reasoning in Chapter Two, the foregoing argument directly applies to the situation of employees in a large private business corporation. Mill contends here that without freedom of expression a person's individuality remains uncultivated; and from both an individual and a social perspective, the development of this trait should be accorded primary importance. Now although the above observations suggest the undesirability of governmental interference with the expression of beliefs and attitudes, they also constitute a strong argument against anything else that undermines the development of individuality. Mill's argument in Chapter Two turns specifically upon the evil that results from according a single person or group the power to regulate expression. By contrast, the undesirable condition associated with a denial of free expression to which Mill subsequently calls our attention—that is, the stifling of individuality—can obtain when coercive interference with the expression of beliefs and attitudes stems from a multitude of independent sources. Accordingly, that restrictions upon expression in the corporate workplace tend to have precisely the above effect would appear to be a compelling ground for holding they should not exist.[18]

It is important to note that Mill's arguments in *On Liberty* do not purport to establish the unjustifiability of any kind of governmental restriction upon expression. Instead, as has been acutely noted by Thomas Scanlon, they are best thought of as directed primarily against *paternalistic* interferences with the expression of thought.[19] Mill's conclusion should be understood as the claim that it is never justifiable for authorities to interfere with the expression of a given thought simply on the ground that such interference is necessary to prevent either (a) harm to certain individuals that consists in their coming to have false beliefs as a result of the expression of that thought, or (b) harm that is the consequence of certain acts that people perform because the thought in question caused them to believe those acts are worth performing. Looked at in the above way, Mill's arguments pertain solely to governmental interference with the expression of beliefs and attitudes based upon their content.

Holding that freedom of expression should exist in the corporate workplace thus commits one to the view that content-based restrictions upon employee speech are never justifiable. Even if what an employee says disrupts the normal corporate decision-making and administrative routines, this price should be paid in order to foster the development of individuality. By the same token, even though an employee's words can harm the reputation of a product unfairly, this no more justifies prior restraints upon employee speech than the possibility that what someone says may result in unfair rejection by the public of a particular governmental policy justifies imposing prior restraints upon undividual citizens or the press. To be sure, declines in productivity and unfair losses of profits are serious matters. But what makes the principle of freedom of expression significant is precisely that it requires important considerations such as these to be subordinated to the interest in maintaining an open atmosphere for the expression of beliefs and attitudes.

To argue against content restrictions upon expression by corporate employees, however, does not rule out regarding other kinds of restrictions as justified. Indeed, it seems to me that most of the situations in which governmental interferences with expression are generally considered justifiable, and hence not violations of the principle of freedom of expression, have analogs in the corporate employer–employee situation. For example, consider the case of an employee who voices dissent continuously during working hours, haranguing other employees so as to make it impossible for them to carry on their work. Sanctions of some kind or other seem reasonable here. This case, however, appears to fall under a rubric similar to the well-entrenched principle of First Amendment case law, that governmental restrictions having to do with time, place, and manner, rather than content, will be upheld so long as they are reasonable.[20] That is, employees should be able to say anything they want, but not necessarily at any time or place or in any manner they choose. Just as in the realm of First Amendment adjudication, however, restrictions in these regards must not be so arbitrary or vague as to be nothing more than thinly veiled subterfuges for regulating the content of employee speech.[21]

To consider another case, what about the disclosure of trade secrets? The issues here appear to be similar to those that arise in connection with officially classified information. The extensive and complicated governmental system for classifying information that has emerged since World War Two has increasingly come to be viewed as incompatible with the basic principles of a free society.[22] Insofar as the rationale for such a system is simply to "prevent sensitive information from falling into the wrong hands," one can justify classifying virtually anything. The classification of information by governmental bodies may not be completely unjustified from the standpoint of the principle of freedom of expression. Nonetheless, it would seem that a legitimate standard for designating material as classified, at the very least, must impose strict limitations as to scope and duration.[23] An analogous proposition appears to hold in the corporate realm. Perhaps some restrictions upon employees from disclosing corporate secrets are consistent with the principle of freedom of expression. The only credible examples I can conceive of, however, would pertain to such matters as the particular figure to be bid on a government contract, the precise formula for a chemical product about to be submitted for a patent, and so forth. In these cases it seems possible to frame relatively narrow restrictions upon expression that would protect the interests of corporate employers without by implication according these employers an unlimited authority to control the content of employee speech subject to no scope or duration restrictions.

Some restrictions upon employee speech thus can be justified. The important point to emphasize, however, is that if the foregoing analysis has merit, then employees should not be prevented or deterred from expressing themselves for reasons having to do with the content of their beliefs and attitudes. The cultivation of individuality fostered by freedom of expression counts for more than almost anything else over the long run. It thus counts for more than the interests that may be compromised by opening up the atmosphere in corporations.

Notes and References

[1]Louis V. McIntire and Marion B. McIntire, *Scientists and Engineers: The Professionals Who Are Not* (Lafayette, La.: Arcola Pub. Co., 1971).

[2]Nicolas Wade, ''Protection Sought for Satirists and Whistleblowers'' *Science* 182 (Dec. 7, 1973), pp. 1002–1003.

[3]A voluminous literature has emerged over the past decade or so with respect to these issues. Among the noteworthy contributions are David Ewing, *Freedom Inside The Organization* (New York: Dutton, 1977); Christopher Stone, *Where The Law Ends* (New York: Harper and Row, 1975); Lawrence E. Blades, ''Employment at Will vs Individual Liberty: On Limiting the Abusive Exercise Of Employment Power'' *Columbia Law Review* 67 (1967), 1404–1435; Phillip I. Blumberg, ''Corporate Responsibility and the Employee's Duty of Loyalty: A Preliminary Inquiry,'' *Oklahoma Law Review* 24 (1971), 279–291; Clyde W. Summers, ''Individual Protection Against Unjust Dismissal: Time For A Statute'' *Virginia Law Review* 62 (1976), 481–532.

[4]Most public sector employees either fall under civil service or are protected by the decision of the United States Supreme Court in Pickering v. Board of Education 391 US 563 (1968) which affirmed very substantial rights to free expression of public employees. As a practical matter, however, the plight of employees in the public sector often does not differ greatly from that of employees in private industry. *See,* United

States. Congress. Senate Committee on Governmental Affairs. *The Whistleblowers* Committee Print. Ninety fifth Congress, second session. US Govt. Print. Off. 1978.

[5]*Ewing,* cited at footnote 3, p. 95.

[6]John J. O'Connor, *New York Times,* Dec. 6, 1973.

[7]For example, the great defenses of freedom of thought one finds in the writings of Milton, Spinoza, Thomas Paine, Voltaire, and John Stuart Mill all take it to be only infringeable by the state. One finds such a presupposition also in the writings of major twentieth century defenders of freedom of expression such as Oliver Wendell Holmes and Alexander Meiklejohn.

[8]See Ewing at pp. 3–29 and Ralph Nader, Mark Green, and Joel Seligman, *Taming the Giant Corporation* (New York: Norton, 1976), pp. 15–32.

[9]One might object that the entire subject of freedom of expression for corporate employees can be disposed of in short order if one simply notes that such employees voluntarily alienate their liberty by accepting jobs with corporations. This view of the matter, however, will not do. One's decision to put up with coercive circumstances in the workplace can only be thought of as fully voluntary to the extent that real options exist for doing otherwise. But in an increasingly bureaucratized society, this is not the case for most people. The view that a corporate employee voluntarily alienates his or her liberty by going to work thus fails for much the same reason as does the thesis, shared by Locke and Plato among others, that one tacitly consents to obey the laws in one's society by not leaving.

[10]*Stone,* cited at footnote 3, p. 213.

[11]*Ewing,* pp. 97–98.

[12]For a general discussion of citizen's arrests, see William F. Prosser, *Law of Torts,* (St. Paul, MN: West Publishing Co., 1971), pp. 42–49.

[13]In this regard the following study is interesting—James Olson, "Engineer Attitudes Toward Professionalism, Employment, and Social Responsibility," *Professional Engineer* 42 (August, 1972), 30–32.

[14]See Ronald Dworkin, "Taking Rights Seriously" in *Taking Rights Seriously* (Cambridge, MA.: Harvard University Press, 1977), p. 184.

[15]John Stuart Mill, *On Liberty* (Currin v Shields, ed.), Library of Liberal Arts Edition, pp. 71–72.

[16]*Op. cit.,* pp. 76–77.

[17]The foregoing interpretation of Mill's argument in chapter three of *On Liberty* departs admittedly from the received view. I have defended this interpretation in my article entitled "Mill's Conception of Individuality" *Social Theory and Practice* 4 (1977), 167–182. I also suggested in an article entitled "A Theory of Personal Autonomy," *Ethics* 86 (1975), 30–48 that Dewey defends the principle of freedom of expression in a manner similar to Mill's approach in chapter three.

[18]William H. Whyte's classic portrait of corporate employees in *The Organization Man* (New York: Simon and Schuster, 1956) provides a compelling illustration of the diverse ways in which corporate life dampens the individual spirit.

[19]See "A Theory of Freedom of Expression," 1 *Philosophy and Public Affairs* (1972), 204–226.

[20]A good review of the pertinent cases in this regard can be found in Gerald Gunther, *Individual Rights in Constitutional Law* (Mineola, NY: The Foundation Press, 1976), 740–804.

[21]In this regard see Lovell v. Griffin 303 US 444 (1938), Schneider v. State 308 US 147 (1938), Hague v. CIO 307 US 496 (1939), Cox v. New Hampshire 312 US 569

(1941), Saia v. New York 334 US 558 (1948), Kunz v. New York 340 US 290 (1951), and Cohen v. California 403 US 15 (1971).

[22]See Benedict Karl Zobrist II, "Reform in the Classification and Declassification of National Security Information: Nixon Executive Privilege Order 11652," *Iowa Law Review* 59 (1973), 110–143.

[23]See Executive Order 11652: Classification of National Security Information and Material, *Federal Register* 37 No. 4 (March 10, 1972).

Whistleblowing: Its Nature and Justification

Gene G. James

Introduction

Whistleblowing has increased significantly in America during the last two decades. Like blowing a whistle to call attention to a thief, whistleblowing is an attempt by a member or former member of an organization to bring illegal or socially harmful activities of the organization to the attention of the public. This may be done openly or anonymously and may involve any kind of organization, although business corporations and government agencies are most frequently involved. It may also require the whistleblower to violate laws or rules, such as national security regulations, that prohibit the release of certain information. However, because whistleblowing involving national security raises a number of issues not raised by other types, the present discussion is restricted to situations involving business corporations and government agencies concerned with domestic matters.

Deterrences to Whistleblowing

It is no accident that whistleblowing gained prominence during the last two decades, which have been a period of great government and corporate wrongdoing. The Viet Nam war, Watergate, illicit activities by intelligence agencies both at home and abroad, the manufacture and sale of defective and unsafe products, misleading and fraudulent advertising, pollution of the environment, depletion of scarce natural resources, illegal bribes and campaign contributions, and attempts by corporations to influence political activities in third world nations are only some of the events occurring during this period. Viewed in this perspective, it is surprising that more whistleblowing has not occurred. Yet few employees of organizations involved in wrongdoing have spoken out in protest. Why are such people the first to know but usually the last to speak out?

The reason most often given for the relative infrequency of whistleblowing is loyalty to the organization. I do not doubt that this is sometimes a deterrent to whistleblowing. Daniel Ellsberg, e.g., mentions it as the main obstacle he had to overcome in deciding to make the Pentagon Papers public.[1] But by far the greatest deterrent, in my opinion, is self-interest. People are afraid that they will lose their jobs, be demoted, suspended, transferred, given less interesting or more demanding work, fail to obtain a bonus, salary increase, promotion, and so on. This deterrent alone is sufficient to keep most people from speaking out even when they see great wrongdoing going on around them.

Fear of personal retaliation is another deterrent. Since whistleblowers seem to renounce loyalty to the organization and to threaten the self-interest of fellow employees, they are almost certain to be attacked in a variety of ways. In addition to such charges as they are unqualified to judge, are misinformed, and do not have all the facts, they are likely to be said to be traitors, squealers, and so forth. They may also be said to be disgruntled, known troublemakers, people who make an issue out of nothing, self-seeking, and doing it for the publicity. Their veracity, life styles, sex lives, and mental stability may also be questioned. Most of these accusations, of course, have nothing to do with the issues raised by whistleblowers. As Dr. John Goffman, who blew the whistle on the AEC for inadequate radiation standards, said of his critics, they "attack my style, my emotion, my sanity, my loyalty, my public forums, my motives. Everything except the issue."[2] Abuse of their families, physical assaults, and even murder are not unknown as retaliation to whistleblowers.

The charge that they are self-seeking or acting for the publicity is one that bothers many whistleblowers. Although whistleblowing may be anonymous, if it is to be effective it frequently requires not only that the whistleblower reveal his or her identity, but also that he or she seek ways of publicizing the wrongdoing. Because this may make the whistleblower appear a self-appointed messiah, it prevents some people from speaking out. Whistleblowing may also appear, or be claimed to be, politically motivated when it is not.

Since whistleblowing may require one to do something illegal, such as copy confidential records, threat of prosecution and prison may be additional deterrents.

Agency Law and Whistleblowing

Not only laws that forbid the release of information, but agency law that governs the obligations of employees to employers seems to prohibit whistleblowing.[3] Agency law imposes three primary duties on employees: obedience, loyalty, and confidentiality. These may be summed up by saying that in general employees are expected to obey all reasonable directives of their employers, to avoid any economic activities detrimental to their employers, and to keep confidential any information learned through their employment that might either harm their employer or that the employer might not want revealed. This last duty holds even after the employee no longer works for the employer. However, all three duties are qualified in certain respects. For example, although the employee is under an obligation to not start a competing business, he or she does have the right to advocate passage of laws and regulations that adversely affect the employer's business. And although the employee has a general obligation of confidentiality, this obligation does not hold if he or she has knowledge that the employer has committed, or is about to commit, a crime. Finally, in carrying out the duty of obeying all reasonable directives, the employee is given the discretion to consult codes of business and professional ethics in deciding what is and is not reasonable.

One problem with the law of agency is that there are no provisions in it to penalize an employer who harasses or fires employees for doing any of the things the law permits them to do. Thus, employees who advocate passage of laws that adversely affect their employers, who report or testify regarding a crime, or who refuse to obey a directive they consider illegal or immoral, are likely to be fired. Employees have even been fired on the last day before their pension would become effective after thirty years of work, or for testifying under subpoena against their employers, without the courts doing anything to aid them. Agency law in effect presupposes an absolute right of employers to dismiss employees at will. Unless there are statutes or contractual agreements to the contrary, an employer may dismiss an employee at any time for any reason, or even for no reason, without being accountable at law. This doctrine, which is an integral part of contract law, goes all the way back to the code of Hammurabi in 632 BC, which stated that an employer could staff the workforce with whomever that employer wished. It was also influenced by Roman law that referred to employers and employees as "masters" and "servants," and by Adam Smith's notion of freedom of contract according to which employers and employees freely enter into the employment contract, so that either has the right to terminate it at will. Philip Blumberg, Dean of the

School of Law at the University of Connecticut, sums up the current status of the right of employers to discharge as follows: "Over the years, this right of discharge has been increasingly restricted by statute and by collective bargaining agreements, but the basic principle of the employer's legal right to discharge, although challenged on the theoretical level, is still unimpaired."[4] The full significance of this remark is not apparent until one examines the extent to which existing statutes and collective bargaining agreements protect whistleblowers. As we shall see below, they provide very little protection. Furthermore, since in the absence of statutes or agreements to the contrary, employers can dismiss employees at will, it is obvious that they can also demote, transfer, suspend, or otherwise retaliate against employees who speak out against, or refuse to participate in, illegal or socially harmful activities.

A second problem with the law of agency is that it seems to put one under an absolute obligation not to disclose any information about one's employer unless one can document that a crime has been, or is about to be, committed. This means that disclosing activities that are harmful to the public, but not presently prohibited by law, can result in one's being prosecuted or sued for damages. As Arthur S. Miller puts it: "The law at present provides very little protection to the person who would blow the whistle; in fact, it is more likely to assess him with criminal or civil penalties."[5] All that the whistleblower has by way of protection is the hope that the judge will be lenient, or that there will be so great a public outcry that the employer will not proceed against him or her.

Lack of Protection for Whistleblowers

There are some laws that encourage or protect whistleblowing. The Refuse Act of 1899 gives anyone who reports pollution one-half of any fine that is assessed. Federal tax laws provide for the Secretary of the Treasury to pay a reward for information about violations of the Internal Revenue Code. The Commissioner of Narcotics is similarly authorized to pay a reward for information about contraband narcotics. The Federal Fair Labor Standards Act prohibits discharge of employees who complain or testify about violations of federal wage and hours laws. The Coal Mine Safety Act and the Water Pollution Control Act have similar clauses. And the Occupational Safety and Health Act prohibits discrimination against, or discharge of, employees who report violations of the act.

The main problem with all these laws, however, is that they must be enforced to be effective. The Refuse Act of 1899 for example, was not enforced prior to 1969, and fines imposed since then have been minimal. A study of the enforcement of the Occupational Safety and Health Act in 1976 by Morton Corn,[6] then an Assistant Secretary of Labor, showed that there were 700 complaints in FY 1975 and 1600 in FY 1976 by employees who claimed they were discharged or discriminated against because they had reported a violation of the act. Only about 20% of these complaints were judged valid by OSHA investigators. More than half of these, that is to say, fewer than three hundred, were settled out of court. The remaining complaints were either dropped or taken to court. Of the 60 cases taken to court at the time of Corn's report in November 1976, one had been won, eight lost, and the others were still pending. Hardly a record to encourage further complaints.

What help can whistleblowers who belong to a union expect from it? In some cases unions have intervened to keep whistleblowers from being fired or to help them gain reinstatement. But for the most part they have restricted themselves to economic issues, not speaking out on behalf of free speech for their members. Also, some unions are as bad offenders as any corporation. David Ewing has stated this problem well. "While many unions are run by energetic, capable, and high-minded officials, other unions seem to be as despotic and corrupt as the worst corporate management teams. Run by mossbacks who couldn't care less about ideals like due process, these unions are not likely to feed a hawk that may come to prey in their own barnyard."[7]

The record of professional societies is not much better. Despite the fact that the code of ethics of nearly every profession requires the professional to place his or her duty to the public above duty to an employer, very few professional societies have come to the aid of members who have blown the whistle. However, there are some indications that this is changing. The American Association for the Advancement of Science recently created a standing committee on Scientific Freedom and Responsibility that sponsored a symposium on whistleblowing at the 1978 meeting and is encouraging scientific societies and journals to take a more active role in whistleblowing situations. A subcommittee to review individual cases has also been formed.[8]

Many employees of the federal government are in theory protected from arbitrary treatment by civil service regulations. However, these have provided little protection for whistleblowers in the past. Indeed, the failure of civil service regulations to protect whistleblowers was one of the factors that helped bring about the Civil Service Reform

Act of 1978. This act explicitly prohibits reprisal against employees who release information they believe is evidence of: (a) violation of law, rules, or regulations, (b) abuse of authority, mismanagement, or gross waste of funds, (c) specific and susbstantial danger to public health or safety. The act also sets up mechanisms to enforce its provisions. Unfortunately, it excludes all employees of intelligence agencies, even when the issue involved is not one of national security, except employees of the FBI who are empowered to go to the Attorney General with information about wrongdoing. Although it is too early to determine how vigorously the act will be enforced, it seems on paper to offer a great deal of protection for whistleblowers.

Although state and local laws usually do not offer much protection, thanks to a series of federal court decisions, people who work for state and municipal governments are also better off than people who work for private corporations. In the first of these in 1968 the Supreme Court ordered the reinstatement of a high school teacher named Pickering who had publicly criticized his school board. This was followed by a 1970 district court decision reinstating a Chicago policeman who had accused his superiors of covering up thefts by policemen. In 1971 another teacher who criticized unsafe playground conditions was reinstated, and in 1973 a fireman and a psychiatric nurse who criticized their agencies were reinstated. In all of these decisions, however, a key factor seems to have been that the action of the employee did not disrupt the morale of fellow employees. Also no documents of the organizations were made public. Had either of these factors been different, the decisions would have probably gone the other way.[9]

Given the lack of support whistleblowers have received in the past from the law, unions, professional societies, and government agencies, the fear that one will be harassed or lose one's job for blowing the whistle is well-founded, especially if one works for private industry. Moreover, since whistleblowers are unlikely to be given favorable letters of recommendation, finding another job is not easy. Thus despite some changes for the better, unless there are major changes in agency law, the operation and goals of unions and professional societies, and more effective enforcement of laws protecting government employees, we should not expect whistleblowing to increase significantly in the near future. This means that much organization wrongdoing will go unchecked.

Objections to Whistleblowing

Whistleblowing is not lacking in critics. When Ralph Nader issued a call for more whistleblowing in an article in the *New York Times* in

1971, James M. Roche, Chairman of the Board of General Motors Corporation responded:

> Some of the enemies of business now encourage an employee to be disloyal to the enterprise. They want to create suspicion and disharmony and pry into the proprietary interests of the business. However this is labelled—industrial espionage, whistle blowing or professional responsibility—it is another tactic for spreading disunity and creating conflict.[10]

The premise upon which Roche's remarks seems to be based is that an employee's only obligation is to the company for which he or she works. Thus he sees no difference between industrial espionage—stealing information from one company to benefit another economically—and the disclosure of activities harmful to the public. Both injure the company involved, so both are equally wrong. This position is similar to another held by many businessmen, viz., that the sole obligation of corporate executives is to make a profit for stockholders for whom they serve as agents. This is tantamount to saying that employees of corporations have no obligations to the public. However, this is not true because corporations are chartered by governments with the expectation that they will function in ways that are compatible with the public interest. Whenever they cease to do this, they violate the understanding under which they were chartered and allowed to exist, and may be legitimately penalized or even have their charters revoked. Furthermore, part of the expectation with which corporations are chartered in democratic societies is that not only will they obey the law, but in addition they will not do anything that undermines basic democratic processes. Corporations, that is, are expected to be not only legal persons, but good citizens as well. This does not mean that corporations must donate money to charity or undertake other philanthropic endeavors, although it is admirable if they do. It means rather that the minimum conduct expected of them is that they will make money only in ways that are consistent with the public good. As officers of corporations, it is the obligation of corporate executives to see that this is done. It is only within this framework of expectations that the executive can be said to have an obligation to stockholders to return maximum profit on their investments. It is only within this framework, also, that employees of a corporation have an obligation to obey its directives. This is the reason the law of agency exempts employees from obeying illegal or unethical commands. It is also the reason that there is a significant moral difference between industrial espionage and whistleblowing. The failure of Roche and other corporate officials to realize this, believing instead that their sole obligation is to operate their companies profitably, and that the sole obligation of employees is to obey their directives without question, is one of the central reasons corporate wrongdoing exists and whistleblowing is needed.

Another objection to whistleblowing advanced by some businessmen is that it increases costs, thereby reducing profits and raising prices for consumers. There is no doubt that it has cost companies considerable money to correct situations disclosed by whistleblowers. However, this must be balanced against costs incurred when the public eventually comes to learn, without the aid of whistleblowers, that corporate wrongdoing has taken place. Would Ford Motor Company or Firestone Rubber Company have made less money in the long run had they listened to their engineers who warned that the gas tank of the Ford Pinto and Firestone radial tires were unsafe? I think a good case could be made that they would not. Indeed, if corporate executives were to listen to employees troubled by their companies' practices and products, in many cases they would improve their earnings. So strong, however, is the feeling that employees should obey orders without questioning, that when an oil pipeline salesman for US Steel went over the head of unresponsive supervisors to report defective pipelines to top company officials, he was fired even though the disclosure saved the company thousands of dollars.

I am not arguing that corporate crime never pays, for often it pays quite handsomely. But the fact that there are situations in which corporate crime is more profitable than responsible action, is hardly an argument against whistleblowing. It would be an argument against it only if one were to accept the premise that the sole obligation of corporations is to make as much money as possible by any means whatever. But as we saw above, this premise cannot be defended and its acceptance by corporate executives in fact provides a justification for whistleblowing.

The argument that employees owe total allegiance to the organizations for which they work has also been put forth by people in government. For example, Frederick Malek, former Deputy Secretary of HEW states:

> The employee, whether he is civil service or a political appointee, has not only the right but the obligation to make his views known in the most strenuous way possible to his superiors, and through them, to their superiors. He should try like hell to get his views across and adopted within the organization—but *not* publicly, and only until a decision is reached by those superiors. Once the decision is made, he must do the best he can to live with it and put it into practice. If he finds that he cannot do it, then he ought not to stay with the organization.[11]

And William Rehnquist, Justice of the Supreme Court, says "I think one may fairly generalize that a government employee . . . is seriously restricted in his freedom of speech with respect to any matter for which he has been assigned responsibility."[12]

Malek's argument presupposes that disclosure of wrongdoing to one's superior will be relayed to higher officials. But often it is one's immediate superiors who are responsible for the wrongdoing. Furthermore, even if they are not, there is no guarantee that they will relay one's protest. Malek also assumes that there are always means of protest within organizations and that these function effectively. This too is frequently not the case. For example, Peter Gall who resigned his position as an attorney for the Office of Civil Rights Division of HEW because the Nixon administration was failing to enforce desegregation laws, and who along with a number of colleagues sent a public letter of protest to President Nixon, says in response to Malek that

> . . .as far as I am concerned, his recommended line of action would have been a waste of everyone's time. To begin with, the OCR staff members probably would have made their protest to Secretary Finch if they had felt that Finch's views were being listened to, or acted upon, at the White House . . . .[13]

And in defense of sending the public letter to Nixon he states that:

> A chief reason we decided to flout protocol and make the text of the letter public was that we felt that the only way the President would even become aware of . . . the letter was through publicity. We had answered too many letters—including those bitterly attacking the retreat on segregation—referred unread by the White House . . . to have any illusion about what the fate of our letter would be. In fact, our standing joke . . . was that we would probably be asked to answer it ourselves.[14]

Even when there are effective channels of protest within an organization, there may be situations in which it is justifiable to bypass them. For example, if there is imminent danger to public health or safety, if one is criticizing the overall operation of an agency, or if using standard channels of protest would jeopardize the interests one is trying to protect.

If Justice Rehnquist's remark is meant as a recommendation that people whose responsibility is to protect the health, safety, and rights of the American people should not speak out when they see continued wrongdoing, then it must be said to be grossly immoral. The viewpoint it represents is one that Americans repudiated at Nuremberg. Daniel

Ellsberg's comments on why he finally decided to release the Pentagon Papers put this point well.

> I think the principle of "company loyalty," as emphasized in the indoctrination within any bureaucratic structure, governmental or private, has come to sum up the notion of loyalty for many people. That is not a healthy situation, because the . . . loyalty that a democracy requires to function is a . . . varied set of loyalties which includes loyalty to one's fellow citizens, and certainly loyalty to the Constitution and to the broader institutions of the country. Obviously, these loyalties can come into conflict, and merely mentioning the word "loyalty" doesn't dissolve those dilemmas . . . . The Code of Ethics of Government Service, passed by both the House and Senate, starts with the principle that every employee of the government should put loyalty to the highest moral principles and to country above loyalty to persons, parties, or government department . . . . To believe that the government cannot run unless one puts loyalty to the President above everything else is a formula for a dictatorship, and not for a republic.[15]

Arguments for Whistleblowing

Even some people who are favorable to whistleblowing are afraid that it might become too widespread. For example, Arthur S. Miller writes: "One should be very careful about extending the principle of whistleblowing unduly. Surely it can be carried too far. Surely, too, an employee owes his employer enough loyalty to try to work, first of all, within the organization to attempt to effect change."[16] And Philip Blumberg expresses the fear that "once the duty of loyalty yields to the primacy of what the individual . . . regards as the 'public interest,' the door is open to widespread abuse."[17]

It would be unfortunate if employees were to make public pronouncements every time they thought they saw something wrong within an organization without making sure they have the facts. And employees ought to exhaust all channels of protest within an organization before blowing the whistle, *provided* it is feasible to do so. Indeed, as Ralph Nader and his associates point out, in many cases "going to management first minimizes the risk of retaliatory dismissal, as you may not have to go public with your demands if the corporation or government agency takes action to correct the situation. It may also strengthen your case if you ultimately go outside . . . since the managers are likely to point out any weaknesses in your arguments and any factual deficiencies in your evidence in order to persuade you that there is really no problem.''[18] But this is subject to the qualifications mentioned in connection with Malek's argument.

If it is true, as I argued above, that self-interest and narrow loyalties will always keep the majority of people from speaking out even when they see great wrongdoing going on around them, then the fear that whistleblowing could become so prevalent as to threaten the everyday working of organizations seems groundless. However, Miller's and Blumberg's remarks do call to our attention the fact that whistleblowers have certain obligations. All whistleblowers should ask themselves the following kinds of questions before acting: What exactly is the objectionable practice? What laws are being broken or specific harm done? Do I have adequate and accurate information about the wrongdoing? How could I get additional information? Is it feasible to report the wrongdoing to someone within the organization? Is there a procedure for doing this? What are the results likely to be? Will doing this make it easier or more difficult if I decide to go outside? Will I be violating the law or shirking my duty if I do not report the matter to people outside the organization? If I go outside to whom should I go? Should I do this openly or anonymously? Should I resign and look for another job before doing it? What will be the likely response of those whom I inform? What can I hope to achieve in going outside? What will be the consequences for me, my family, and my friends? What will the consequences of *not* speaking out be for me, my family, and my friends? What will the consequences be for the public? Could I live with my conscience if I do not speak out?

There is one respect in which whistleblowing might be taken too far. However, this requires explanation. Proponents of whistleblowing often write as though it were by definition a morally praiseworthy activity. For example, whistleblowing is defined in the Preface of Nader's book as

> . . .the act of a man or woman who, believing that the public interest overrides the interest of the organization he serves, publicly ''blows the whistle'' if the organization is involved in corrupt, illegal, fraudulent, or harmful activity . . . .[19]

And Charles Taylor and Peter Branch, defending whistleblowers against the charge that they are traitors, say:

> The traitor was always hated, because he was the enemy, but there was a special edge on the scorn that historically made traitors hated everywhere . . . . For the traitor was excoriated as a person without honor of any kind, who, among people willing to die for cause or principle . . . could shuffle back and forth between

> opposing camps, sniffing for the highest bidder, unmoved by higher loyalty or human bond . . . . Benedict Arnold was hated for defecting . . . but he was despised as a real traitor because people found out that he had bargained at length over the pension he would receive—and even the number of calico dresses his wife would obtain—for switching sides.
>
> The whistle-blowers have actually reversed the operation of the classical traitor, as they have usually been the *only* people in their organizations taking a stand on some kind of ideal.[20]

They go on to argue that whistleblowers have been so successful in winning public admiration that "looking at the problem through the eyes of future whistle-blowers, the dilemmas are likely to center not on the morality of proposed actions but on their utility."[21]

The problem with both this definition and defense of whistleblowing is that they fuse motives with goals. That is, they seem to take for granted that because the whistleblower discloses wrongdoing, his or her motives must be praiseworthy. In fact, people may blow the whistle for a variety of reasons: to seek revenge, gain prosecutorial immunity, attract publicity, and so on. Furthermore, from the standpoint of immediate public good, the motives of whistleblowers are unimportant. All that matters is whether the situation is as the whistleblower describes it. That is, does a situation exist which is harmful to the public?

However, consideration of the motives of whistleblowers does raise an interesting problem. To what extent should there be laws that encourage whistleblowing for self-interested reasons? Should there, e.g., be more laws like the federal tax law that furthers disclosure of tax violations by paying informers part of any money that is collected? Do not laws such as these bring about a situation that encourages suspicion, revenge, and profit seeking among citizens? And would not a society that relied heavily on this type of law be "taking whistleblowing too far?" I think that the answer to the last two questions must be affirmative. This does not mean, however, that society should never adopt laws that encourage whistleblowing for base motives. The extent to which a given law furthers spiteful behavior among citizens must be balanced against the amount of good it produces. In some cases, although I do not think in very many, the good clearly outweights the harm of furthering vengeful behavior.

It would seem that anyone who was in favor of whistleblowing would also be in favor of laws protecting whistleblowers from arbitrary discharge from their jobs. However, Peters and Branch argue that with one exception

> . . . freedom to hire and fire without red tape is essential to good government and . . . potential whistle-blowers should not be promised a world free of risks. Of course, our society should offer everyone a cushion against catastrophic job loss in the form of a decent guaranteed annual income and free health care. Beyond that we resist depriving life of adventure.[22]

The one exception they allow is that of people in situations like that of Dr. Jacqueline Verrett, the FDA employee who informed the press that her superiors were distorting the results of experiments she had performed, which showed that cyclamates cause growth deformities in chicken embryos. All other whistleblowers, they say, even those such as Daniel Ellsberg whom they admire greatly, should be subject to being fired. But, exactly why people in situations such as Dr. Verrett's should be treated as exceptions is not made clear. Perhaps they consider threats to people's health more important than other threats to their well-being. Nor do they give any argument to justify why not depriving life of adventure is more important than preventing the discharge of people they claim are usually the most conscientious employees of organizations. Their position also seems to be inconsistent since they say elsewhere that whistleblowing "should be encouraged, even by . . . employers themselves."[23]

The reason they arrive at such a contradictory position, in my opinion, is that they too fear that whistleblowing might get out of hand. Shortly after making the foregoing remark they say:

> There should be more protection for the whistle-blowers who prove right . . . without making whistle-blowing an automatic free ride. The risk must be preserved, for otherwise whistle-blowing would become banal, the country would be inundated with exposures, and the good cases would become uselessly lost in a sea of bad ones.[24]

This is to abandon the view that all whistleblowers should be fired, in favor of the one that people whose claims turn out to be false should be fired. But should all whistleblowers be fired whose claims turn out to be false? If so, what about those whose claims are partially true and partially false? The position maintained by Peters and Branch is a variation of the view that speech should be protected only when what is said is true. I doubt that they would accept this view if applied to the press. As working journalists they are more likely to believe that the press should be penalized only when it can be shown that false statements were made with malicious intent. Why should it be different in the case of whistleblowers? To believe that it should is not only to deprive life of adventure, it is to actively discourage whistleblowing and to allow corporate and governmental wrongdoing to flourish.

Rosemary Chalk, Staff Officer for the American Association for the Advancement of Science, arguing for greater involvement of scientific societies in whistleblowing, is correct, in my opinion, when she says that: "It should not be necessary for the whistle-blower to be 100% correct in order to gain support from his or her professional colleagues. The basis for scientific society involvement should not rest exclusively on whether the whistle-blower is right or wrong, but rather on whether the issue . . . is important in terms of its effect on the public interest."[25] Philip Blumberg is also correct when he says: "The public interest in the free discussion of ideas does not rest on the validity of the point of view expressed. Where dissent involves *no unauthorized disclosure,* the cost of sanctioning such conduct is low and of prohibiting it, high."[26] But unlike Blumberg, I believe that the area of authorized disclosure should be as wide as possible. This means that the law of agency should be superseded by federal legislation that would prevent employers from discharging or otherwise penalizing whistleblowers unless it can be proven both that their claims are false and were made with malicious intent. However, the situation must be one in which the public interest is at stake. Disclosure of trade secrets, customer lists, plans for marketing, personnel records, and so on should not be protected unless releasing them was necessary for the whistleblowing and the damage resulting is clearly outweighed by benefits to the public.

Many people are opposed to legislation protecting whistleblowers because they believe it unwarranted interference with freedom of contract. However, the traditional doctrine of freedom of contract rested on an assumption of equality between employers and employees that no longer exists. It is far easier today for employers to find another employee than for employees to find a new job. Furthermore, as Lawrence E. Blades has pointed out, the freedom of the individual to terminate his or her employment is more important than the freedom of employers to hire and fire at will.[27] The rights of the individual to dispose of one's labor as one wishes and to be protected from retaliation in exercising one's civil rights are fundamental to the existence of democratic societies. The courts have recognized this to some extent by limiting the right of employers to dismiss employees for union activities. The current practice of allowing employers to dismiss employees who are performing their jobs competently and whose sole "offense" is disclosure of situations harmful to the public is, therefore, one that ought to be abolished.

If legislation to protect whistleblowers is effective, it should provide for punitive damages against employers when it can be proven that they dismissed or penalized employees for justified whistleblowing. But how is the employee to show this? Both Blades and Ewing believe that the burden of proof should be on the employee. Blades writes:

> Ordinarily, when both sides present equally credible versions of the facts, the plaintiff will have failed to carry his burden. However, there is the danger that the average jury will identify with . . . the employee. This . . . could give rise to vexatious lawsuits . . . .
>
> Certainly, the employee should not be allowed to shift to the employer the burden of showing that the discharge was motivated by good cause by proving only that he capably performed the duties required by his job and was discharged for no apparent reason . . . . The employee should be required . . . to prove by affirmative and substantial evidence that his discharge was actuated by reasons violative of his personal freedom or integrity.[28]

Blades and Ewing qualify their position by suggesting that longevity of service be taken as evidence that an employee was doing an effective job. Blades says "in cases like *Mims,* where the discharged employee had served the employer for 17 years as a branch manager and 32 years in all, a jury would probably be quite justified in finding little merit in an explanation that the plaintiff was fired for 'chronic' inefficiency and incompetence."[29] And Ewing states: "The longer the ex-employee was on the job, the better his or her case . . . . If management was doing its job, such employees [who have served eight or more years] must have been working competently or they would not have stayed on the payroll all that time."[30]

The problem with these remarks is that if the burden of proof is on the employee to show by "affirmative and substantial" evidence that the discharge was for reasons that violate his or her personal freedom or integrity, and the employer is under no obligation to show that the discharge was for good cause, then establishing that an employee has competently performed a job for a number of years would be irrelevant to the issue at dispute. Showing this would be relevant only if the employer were under an obligation to show that he or she had not performed competently. And in cases involving discrimination against, or the firing of, whistleblowers, this is exactly what employers should be required to do. To require that employees show by affirmative and substantial evidence that they were not fired for good reason, when employers are under no obligation to demonstrate why they were fired, is to require a degree of evidence impossible to fulfill. Such a law would offer no protection to whistleblowers and is in fact a reversion to the doctrine of the absolute right of employers to discharge at will that any laws protecting whistleblowers should be designed to overcome.

The fear that unless the burden of proof is placed on the whistleblower, vexatious lawsuits will result is unrealistic for two reasons. First, as I argued above, self-interest will always keep the majority of people from engaging in whistleblowing. Second, only employees who could show that an act of whistleblowing preceded their being dismissed or penalized would be able to seek redress under the law.

Other Needed Reforms

Laws preventing whistleblowers from being dismissed or penalized are, of course, only one way of dealing with corporate and government wrongdoing. Laws and other measures aimed at changing the nature of organizations to prevent wrongdoing and encourage whistleblowing are equally important. Changing the role of corporate directors, appointing ombudspersons and high level executives to review charges of wrongdoing, requiring that certain types of information be compiled and retained, and reforming regulatory agencies are some of the proposals that have been advanced for doing this. Unions and professional societies also need to act to further the rights of their members and protect the public good. Professional societies, e.g., should reformulate their codes of ethics to make obligations to the public more central, investigate violations of member's rights, provide advice and legal aid if needed, censure organizations found guilty, attempt to secure legislation protecting member's rights, and in some cases set up central pension funds to free members from undue dependence on the organization for which they work.

Since my primary purposes were to clarify the nature of whistleblowing, defend it against criticisms, and show its importance for democratic societies, detailed discussion of the topics mentioned in this section is beyond the scope of this paper. Whistleblowing is not a means of eliminating all organizational wrongdoing. But in conjunction with other measures to insure that organizations act responsibly, it can be an important factor in maintaining democratic freedom.

Notes and References

[1]See Charles Peters and Taylor Branch, *Blowing the Whistle: Dissent in the Public Interest* (New York: Praeger, 1972), Chapter 16.

[2]Quoted in Ralph Nader, Peter J. Petkas, and Kate Blackwell, *Whistle Blowing* (New York: Grossman, 1972), 72.

[3]For a discussion of agency law and its relation to whistleblowers see Lawrence E. Blades "Employment at Will vs. Individual Freedom: On Limiting the Abusive Exercise of Employer Power," *Columbia Law Review* 67 (1967), and Philip Blumberg, "Corporate Responsibility and the Employee's Duty of Loyalty and Obedience: A Preliminary Inquiry," *Oklahoma Law Review* 24 (1971), reprinted in part in Tom L. Beauchamp and Norman E. Bowie, *Ethical Theory and Business* (New York: Prentice-Hall, 1979); and Clyde W. Summers, "Individual Protection Against Unjust Dismissal: Time for a Statute," *Virginia Law Review* 62 (1976). See also Nader, *op. cit.*, and David W. Ewing, *Freedom Inside the Organization* (New York: Dutton, 1977).

[4]Blumberg, *op. cit.*, in Beauchamp and Bowie, p. 311.

[5]Arthur S. Miller, "Whistle Blowing and the Law," in Nader, *op. cit.*, p. 25.

[6]Corn's report is discussed by Frank von Hipple in "Professional Freedom and Responsibility: The Role of the Professional Society," in the *Newsletter on Science, Technology and Human Values* (Number 22, January 1978), pp. 37–42.

[7]Ewing, *op. cit.*, pp. 165–66.

[8]Discussion of the role professional societies have played in whistleblowing can be found in Nader, *op. cit.*, Von Hipple, *op. cit.*, and in Rosemary Chalk, "Scientific Involvement in Whistle Blowing" in the *Newsletter on Science, Technology and Human Values, op. cit.*, pp. 47–51.

[9]These decisions are discussed by Ewing, *op. cit.*, Chapter 6.

[10]Quoted in Blumberg, *op. cit.*, p. 305.

[11]Quoted in Peters and Branch, *op. cit.*, pp. 178–179.

[12]*Ibid*, pp. x–xi.

[13]*Ibid*, p. 179.

[14]*Ibid*, p. 178.

[15]*Ibid*, p. 269.

[16]Miller *op. cit.*, p. 30.

[17]Blumberg, *op. cit.*, p. 313.

[18]Nader, *op. cit.*, pp. 230–231.

[19]*Ibid*, p. vii.

[20]Peters and Branch, *op. cit.*, p. 288.

[21]*Ibid*, p. 290.

[22]Peters and Branch, *op. cit.*, p. xi.

[23]*Ibid*, p. 298.

[24]*Ibid*.

[25]Chalk, *op. cit.*, p. 50.

[26]Philip Blumberg "Commentary on 'Professional Freedom and Responsibility: the Role of the Professional Society' " in the *Newsletter on Science, Technology and Human Values, op. cit.*, p. 45.

[27]Blades, *op. cit.*

[28]*Ibid*, pp. 1425–1426.

[29]*Ibid*, pp. 1428–1429.

[30]Ewing, *op. cit.*, p. 202.

On the Rights of Professionals

Albert Flores

Introduction

A great deal of the current literature in the field of "professional ethics" concerns itself either with the duties and professional responsibilities that professionals owe to clients, colleagues, employers, and society, or with case histories of their failures to adequately satisfy these duties and responsibilities. From the point of view of the professions this is, of course, understandable given the importance that they place on practicing according to the standards laid down in their codes of ethics. It is reasonable too, from the wider perspective of the public interest. Professional authority, if abused, can adversely affect the interests of society and this may undermine public confidence in a profession and its practitioners. Thus, it is important that professionals acknowledge their duties and proper that much of the discussion of professional ethics focus on professional responsibilities.

Unfortunately, although these are issues that deserve our careful attention, it is apparent that we have ignored a topic intimately connected with the issue of professional responsibility, and one perhaps equal in importance to the public's interest, *viz.* the *rights* that individuals possess as a consequence of their *professional* status.

The "rights of professionals" are seldom discussed and very rarely formally recognized. Consider, for example, the fact that in the codes of ethics of the major engineering societies, there is no mention of the rights of engineers, either in relation to their clients, their colleagues, or the public. This is generally true for most of the other major professions as well.[1] Whereas ethical codes unequivocally instruct professionals in what they should or should not do, nothing is said about those things to which they are entitled as professionals. One comes away with the impression that professional ethics is concerned simply with prohibitions and obligations, and not with anything positive, such as what professionals are justified in doing without interference or jeopardy because it is their *professional right*.

From the perspective of ethics this is a distortion, for the rights of individuals are as morally significant as their duties, responsibilities, or obligations. When our rights are violated, our indignation is no less strong than when we criticize others for failing in their obligations. Indeed, an appeal to a right is often the major premise upon which many moral arguments depend, as when we appeal, for example, to the right to life, the right to privacy, or the right to work.

But what does it mean to say that professionals have rights and how are professional rights different from the other rights we have as individuals? It is the purpose of this paper to try to provide answers to these questions. In the next section, I analyze the concept of "right" and lay out some of its logical implications. With this task completed, I attempt to sketch out a general theory of professional rights and its attendant justification. In the last section, I show why professional rights are important by examining the case of professional engineers.

The Logic of Rights

What do we mean when we say that individuals have rights: for example, the right to speak freely or the right to be treated with respect?[2] In general, we normally mean that they are entitled to expect that *others* will not interfere with them when they act. For example, when we say that individuals have a right to publicly express an opinion on the development of nuclear power as an energy source, we mean that it would be wrong for others to prohibit or interfere with their doing so. A right is a capacity to act that others are duty-bound to respect. In other words, to have a right is to be in a position to make a claim against the behavior of others.

The nature of this claim is in the form of an obligation or duty that others owe to us as right holders. Jeremy Bentham argues that

> . . . it is not possible to create rights which are not founded on obligations. How can a *right* of property in land be conferred upon me? It is by imposing upon everybody else the obligation of not touching its productions. How can I possess the right of going into all the streets of a city? It is because there exists no obligation that hinders me, and because everybody is bound by an obligation not to hinder me.[3]

Rights are, therefore, justified by the obligations that other individuals have with respect to us. These obligations are of two kinds: *positive* obligations, in the sense that a right holder profits or benefits from the actions of others because they are duty-bound to give us something; and *negative* obligations, in that others are required to restrain themselves from interfering with us when we act. The right to do something, then, implies that there is no existing obligation that prohibits our acting in a particular way, but more importantly, that others are obliged to us either because they owe us something we are entitled to, or because it would be wrong for them to interfere with our performance of this action.

However, to say that individuals have *rights* is not to say that how we exercise these rights is "right," in the sense that it is the correct thing to do. We must distinguish between the exercise of a right and its content; that is, between the rights of the individual and the correctness of what the individual does. For instance, although engineers may have a right to express their views on nuclear power, it is obvious that not all of these views will be correct or logically coherent, as is demonstrated by the range of different and contradictory opinions that engineers have held on this issue. Furthermore, we may have a right to speak freely, but it would be wrong, for example, to spread rumors about colleagues, even though doing so is an instance of exercising the right to free speech. In other cases, where the exercise of our right to speak freely does not involve doing something wrong, such as lying, it may still be the wrong thing to do from the point of view of ethics. We may have a right to express an opinion on the airworthiness of the DC-10, for example, but it would be wrong to exercise this right if our opinions are based on faulty, misleading, or inadequate information. If we are poorly informed, then we should not speak, even though we have a right to speak. This is especially true in those circumstances where our opinion will be taken seriously or may have significant implications, for example, when testifying in a court of law or before a legislative committee. In short, though we have the right to free speech, how and when we speak is a matter that is susceptible to both logical and moral evaluation, and this holds true for all the other rights we may possess as well. Thus to say that individuals have rights or a right to do something does not mean that what they do in exercising these rights is logically correct or ethically justified.

This conclusion suggests a related and equally important fact about rights. The possession of a right normally entails that individuals have discretion as to whether or not to exercise their rights. Professionals have, for example, a right to adequate compensation for work performed. But they may choose to waive this right when, for instance, they donate their time and expertise for charitable causes. Joel Feinberg argues that recognizing the option *not* to exercise a right makes possible a number of charitable or "supererogatory" acts that are morally praiseworthy.[4] Thus, though we may be entitled to something as a right, in some cases we may choose not to exercise a right. There are, however, important exceptions. Some rights cannot be waived if we are to maintain our dignity as human beings and our integrity and self-esteem as professionals. Historically, governments have been most responsible for taking advantage of our failure to exercise certain rights, but with the pervasive growth and influence of organizations and institutions employing large numbers of individuals, including professionals, the threat to individual rights from this source is even more insidious.[5] We shall return to this point in the last section.

To summarize the logic of rights: when we say that individuals have rights we are asserting that they are entitled to benefit or profit from the behavior of others against whom the right is claimed. A right is an *interest* deserving special protection that may be derived from law, socially established rules, or by the fact of one's status, (e.g., human being, parent, citizen, creditor, professional, and so on). It imposes obligations on others either to give to the right holders something they are warranted in claiming as their own, or to refrain from interfering or prohibiting them from performing acts to which they are entitled as their right. When individuals demand their rights, they are seeking satisfaction from those who have obligations to them for which they have a legitimate claim. Finally, not all the actions we may perform as a right are necessarily morally right or logically correct. Rights authorize our action, but do not justify every act that falls under them.

Rights of Professionals

If this is what we mean when we say that individuals have rights, what then are *professional* rights?

The answer I propose is an instance of a more general theory of natural rights, in that it appeals to a common notion of what it means and why it is important to have rights. It assumes that there is an intimate connection between rights and having a valued status or characteristic, and that rights are a way of protecting certain activities or interests that contribute in some essential way to the existence of this status or characteristic. Thus, just as human rights, like the right to freedom of speech, set off certain activities for special protection from the interference of others or provide for what may be our just due, professional rights will serve a similar purpose.

In order to understand the nature of professional rights, it will be helpful to consider the example of how *legal* rights are established and justified. This is a useful model because there are many similarities between the case for legal rights and professional rights. It illustrates too, some of the properties that make rights important.

Consider again the right of individuals to freedom of speech, which in our society is a legal as well as human right. By granting its members the right to free expression of ideas, a community serves several valued ends, perhaps the most important of which are justice and political harmony. If members of a community are unable to express their ideas and defend their interests, then in such a repressive atmosphere they will not have a fair chance of having these interests considered when decisions affecting them are made by the community. The injustice of such a situation is obvious and the frustration it can foment has been responsible for some of history's most celebrated political upheavals. Moreover, without the freedom to express one's self and in this way pursue one's interests, it is impossible for members of a community to develop capabilities that can contribute to the community's welfare.

To forestall such calamities and in the interests of justice, a community can impose a legal obligation on all its members to respect and refrain from interfering with any individual's freedom to speak. By imposing this negative obligation on its members, a community acknowledges that open expression of ideas is vital to its overall interests, as well as to the interests of its citizens.

We may, however, go beyond this utilitarian argument and further assert that freedom of speech may be one of several rights that individuals must possess if they are to remain *citizens* at all. Simply stated, individuals cannot realistically participate in the political processes of a community unless they have an opportunity to have their opinions known and to judge the opinions of others. These are activities that are an essential part of a citizen's role. In short, free speech is functionally necessary if individuals are to fulfill their role as citizens. Thus legal rights—and the duties they imply—are a community's way of identifying certain activities or interests as things individuals are entitled to have and enjoy without interference or restriction. In general, in order to know which legal rights individuals should possess, we need only specify the activities that are essential to one's status as a citizen.

Now, consider the case for professional rights. The application of professional skills is instrumental in producing goods of significant practical value. Lawyers can, for example, assist individuals in the protection of property and other interests, physicians can bring the injured and diseased back to health, and engineers can develop technologies that address the basic needs of society. The production of these useful goods, like the social goods that freedom of speech promotes, is fostered by respect for the professional judgment and knowledge that underlies the application of these skills. Clients who consistently ignore or subvert their lawyer's or physician's advice are not likely to find justice or be cured, and a public that spurns new technologies will have greater difficulty in satisfying the necessities of life. In short, these and other goods are dependent in great measure upon the willingness of clients to show due respect for professional judgment and expertise. In recognition of this fact, communities have traditionally accorded professionals an exalted social status that has served, in part, to signify the importance it accords professional authority. By imposing on its members a positive obligation to show proper respect for professional authority, a community acknowledges that professionals are entitled to have their professional opinions taken seriously. More generally, professional rights to practice law, dispense drugs and perform surgery, or engage in engineering are exclusively granted to lawyers, physicians, and professional engineers, respectively, and involve imposing special obligations on nonprofessionals to refrain from performing these activities.

In addition, professional rights may be justified by reference to the function, goal, or purpose of a profession. To say professionals have rights *qua* professional implies that certain activities and interests that one may have as a professional are crucial to the satisfaction of one's professional function. For example, lawyers and physicians could not practice unless they have the right to protect privileged information. In order to serve the client's needs, these professionals require sensitive information concernng a client's legal or medical situation that a client would have no good reason to reveal without first obtaining an assurance that it will remain confidential. Such an assurance is worthless if physicians and lawyers are unable to have these confidences respected by those outside the professional–client relationship. Thus, the right of confidentiality not only promotes trust, but it is necessary if these professionals are to fulfill their professionally defined functions. A similar argument may be made for engineers. Engineers are required to "hold paramount the safety, health, and welfare of the pulic in the performance of their professional duties" (ECPD Code of Ethics for Engineers). But in order to satisfy this duty it is necessary that engineers be allowed to judge the technical and design details of products to whose production they contribute. Without this right of review they cannot be expected to fulfill their duties to protect the public from harmful products, materials, or constructions. Thus, activities that promote public safety are matters which identify some of the professional rights of engineers.

Generally, professional rights can be established and justified either by reference to the value of goods that the special protection of certain professional activities contributes to producing, or by reference to those activities that are essential to fulfilling a professions's purpose, goal, or function. In most cases, the goods that respect for professional rights generates are simply a material manifestation of a profession's function. Thus some professional rights may be justified in either of two ways. It should, however, be noted that it is not always easy to identify the specific goods that the existence of a professional right contributes to producing, or to what degree these goods may require the imposition of obligations on nonprofessionals in order to insure their production. Moreover, it is equally difficult to clearly state a profession's function, let alone identify the activities that are essential to its satisfaction. In part, this is a problem concerning competing conceptions of what the *ideal* professional lawyer, physician, engineer, and so on should be, and obviously this will engender considerable debate over the kinds of rights professionals can claim as their professional right. But if we can come to some agreement on the end of a profession and on which activities are crucial to achieving its valued goods, we may want to single these out as activities that professionals are entitled to perform as their right.

One final cautionary note: Some of the rights that are claimed as "professional" rights are in reality *contractual* rights. Those who make a contract voluntarily incur obligations that they owe to the person with whom they have contracted. The benefits resulting to either party from this contract can be claimed as a right, i.e., something to which one is entitled. So, for example, the right of professionals to receive adequate compensation for work they perform is not a "professional" right *per se*. Rather it is grounded in a contract between professional and client. The benefit that a client receives from the exercise of a professional's skills requires appropriate compensation for which the physician, lawyer, engineer, and so on is entitled. Here, a contract establishes the professional's right and imposes on the client an obligation to honor this claim; clearly, one can have this kind of right regardless of one's status. Thus, these and similar rights are not strictly speaking "professional" rights, though professionals can have them.

The Significance of Professional Rights

Problems of identifying and justifying professional rights are worth pursuing only if there are good reasons to acknowledge that professionals have rights. Why is it important that the rights of professionals be recognized? In order to begin to answer this question it may prove useful to examine a case in which professional rights play a significant role.

Consider the following:

> Engineers of Company "A" prepared plans and specifications for machinery to be used in a manufacturing process and Company "A" turned them over to Company "B" for production. The engineers of Company "B" in reviewing the plans and specifications came to the conclusion that they included certain miscalculations and technical deficiencies of a nature that the final product might be unsuitable for the purposes of the ultimate users, and that the equipment, if built according to the original plans and specifications, might endanger the lives of persons in proximity to it. The engineers of Company "B" called the matter to the attention of appropriate official of their employer who, in turn, advised Company "A" of the concern expressed by the engineers of Company "B." Company "A" replied that its engineers felt that the design and specifications for the equipment were adequate and safe and that Company "B" should proceed to build the equipment as designed and specified. The officials of Company "B" instructed its engineers to proceed with the work.[6]

What should the engineers of Company "B" do in this situation?

This is a classic example of a moral dilemma occasioned by the inconsistent demands of conflicting duties. Should these engineers follow their professional opinion and refuse to contribute to the production of what they honestly believe will be unsafe equipment? Or, should they obey their employer's instructions to proceed with production, regardless of their unresolved safety concerns?

The resolution of this dilemma depends a great deal on whether the engineers in this case are justified, and perhaps entitled, to refuse to work on this project because it violates critical professional concerns involving human safety. Insofar as they are *employees,* these professionals are subject to their employer's authority, hence they cannot realistically consider ignoring their employer's wishes without at the same time risking their jobs. From the point of view of the employer's interests this is not unreasonable. But should employed professionals be entitled to withdraw their skills, without fear of reprisal, when these skills may be misused or abused by an employer? Clearly, without some such protection, employed professionals, like the engineers in this case, will likely be forced on such occasions to sacrifice some important professional concerns in order to avoid jeopardizing their career. Because employers can have such a powerful influence over the quality of professional practice, it seems that some limit on the employer's authority should be established in the form of a negative obligation to refrain from interfering with the exercise of professional authority in areas where professionals alone have the competence. In short, professionals should have special protections that are identified and acknowledged as their *professional right* that no employer or other nonprofessional can ignore or violate.

Historically, rights have functioned as a way of protecting individuals and their interests from the abuse of the power and authority that rulers, police, institutions, and so on have over the individual.[7] It is not surprising that, with the growth and concentration of power in these sources since the middle ages, there would develop a need to identify certain interferences in matters of personal sovereignty that no power could justly exercise over an individual. Identifying something as a *right* was thus intended to insulate individuals from the authority of others by imposing on these others obligations either to refrain from interfering or to provide what is their just due. In this way communities have acknowledged the importance of protecting vital individual interests from the tyranny of authority.

The importance of professional rights may be similarly explained. Consider again the profession of engineering. In general, most engineers practice their profession as employees of industry or government. Except for a small minority of self-employed consultants, few engineers enjoy the autonomy that has historically characterized the more traditional professions; consequently, their rights as professionals are easy to ignore. Moreover, many decisions relating to engineering design, manufacture, and construction, highly technical matters that only an engineer may be qualified to assess, are often routinely made by superiors in management who are normally without the necessary expertise to decide such matters. In an organizational context in which one's actions are subject to review by superiors, it is difficult to assert or demand that professional rights be recognized. In many cases, this abridgement of professional authority can adversely affect the public's interests, as the growing concern over product safety so amply demonstrates.[8] Indeed, a profession's failure to acknowledge the importance of professional rights may contribute to a public attitude that can undermine professional authority and that may lead to a loss of respect for professional judgment. And without the autonomy and authority to act on professional judgments dictated by one's skill and conscience, it is doubtful that we can then hold engineers wholly responsible for their activities when these may be harmful to the public's welfare. Thus, just as our natural right to act on our conscience or the right to freedom of speech are examples of rights we cannot choose not to exercise without also abdicating our role as responsible moral agents, engineers cannot afford to waive those rights that are essential to the practice of their profession and remain faithful to the character that defines a *professional* engineer.

Therefore, given the recent trend of increasing institutional control and supervision of the routine activities of most major professions, including medicine and law, practicing professionals need a way of protecting those vital professional interests over which they alone have competence and should have jurisdiction.

If a "professional" is defined as someone who has a skill that serves an essential social need, then in order to meet these needs properly professionals must be free to act on important professional concerns without interference. In an organizational context, however, professional interests and judgments will not necessarily take priority over the interests and needs of the organization. Realistically, this is to be expected, but the effect may be a perversion of professional judgment serious enough to affect a profession's ability to adequately and safely serve society's needs. To avoid this unacceptable consequence, it may be necessary to acknowledge the existence of special rights that organizations are obliged to respect and that employed professionals are entitled to act upon by the mere fact of their professional status.

Acknowledgments

This is a revision of a paper previously published in *Issues in Engineering* (Autumn 1980) entitled "Engineers' Professional Rights." I extend my thanks to Deborah Johnson and Fred Elliston for their helpful criticisms of an earlier draft of this paper, and to Frances Anderson for her assistance in preparing the manuscript. This paper is based upon research supported by a grant from the National Endowment for the Humanities, Grant No. 27912-77-271.

Notes and References

[1]Of the major professions, the only exceptions to this claim are in the American Dental Association's "Principles of Ethics," which mentions in its first section "The right of a dentist to professional status rests on the knowledge, skill and experience with which he serves his patients and society"; and in the American Medical Association's former code it mentions that "a physician may choose whom he will serve," but this code has recently been totally revised.

[2]For brief discussions of various positions on rights see: David Lyons, ed., *Rights,* (Belmont, CA: Wadsworth Publishing Co., 1979); also Richard E. Flatman, *The Practice of Rights* (Cambridge, MA: Cambridge University Press, 1976); and Ronald Dworkin, *Taking Rights Seriously* (Cambridge, MA: Harvard University Press, 1977).

[3]Jeremy Bentham, *A General View of a Complete Code of Law* (1802) reprinted in *Works of Jeremy Bentham,* Vol. III, p. 181.

[4]Joel Feinberg, "The Nature and Value of Rights", reprinted in *Rights, op. cit.,* pp. 78–91.

[5]See *Individual Rights in the Corporation: A Reader on Employee Rights,* Alan F. Westin and Stephan Salisbury, eds. (New York: Pantheon, 1980); David Ewing, *Freedom Inside the Corporation* (New York: Dutton, 1977); and Paul Goodman, *People or Personnel* (New York: Vintage Books, 1968).

[6]Reprinted from *Ethical Problems in Engineering,* Vol. II, Cases, Robert Baum, ed. (Troy, NY: Rensselaer Polytechnic Institute, 1980), p. 13.

[7]Previous to the late middle ages, neither legal nor ethical theory recognized the existence of rights, see: "The Concept of Rights: A Historical Sketch," by Martin P. Golding, in *Bioethics and Human Rights,* E. L. Bandman and B. Bandman, eds. (Boston: Little Brown, 1978), 44–50.

[8].Two recent books that highlight this problem are: John Kolb and Steven S. Ross, *Product Safety and Liability: A Desk Reference,* (New York: McGraw-Hill, 1980); and James F. Thorpe and William H. Middendorf, *What Every Engineer Should Know About Product Liability,* (New York: Dekker, 1979.)

Index

Index